Permaculture

teachers' guide

Edited by Andrew Goldring

WWF

Permaculture
ASSOCIATION

Published by WWF-UK in association with the Permaculture Association and Permanent Publications

WWF-UK registered charity number 1081247
A company limited by guarantee number 4016725
Panda device ©1986 WWF ® WWF registered trademark owner
The Permaculture Association is UK registered charity number 1116699

A catalogue record for this book is available from The British Library

ISBN 1 85850 168 7

Printed and bound in Great Britain by
CPI Antony Rowe, Chippenham and Eastbourne

Designed by WWF Design Team
Mind maps and illustrations by Andrew Goldring from originals by the authors

Warning!

Please note that the sessions contained within this book should be well rehearsed and well prepared before use. This book is not intended to replace the direct sharing of skills between teachers and apprentices, or your own research and preparation.

For up-to-date contact details of all the contributors, details of courses and support for teachers, including teacher training, contact the Permaculture Association:
Write to: BCM Permaculture Association, London, WCIN 3XX
Telephone: 0845 4581805
Email: office@permaculture.org.uk
Internet: www.permaculture.org.uk

Foreword

I can remember being struck by Edward de Bono's assertion in Parallel Thinking that there can be more than one 'right way'. I have subsequently adapted this on many training occasions by asking groups to describe the right way to get from London to Birmingham. Of course, no one ever falls for it, but the point is made that for different needs there are different 'right ways'.

There is an obvious parallel here with the principles that underpin Permaculture. Any approach that is based on a systems design concept will throw up a range of solutions, some of which are more applicable than others, but all with some merit. Permaculture deserves a high profile because its design approach is fundamental in any problem solving situation. Professor John Stewart of the Local Government School of Public Policy, University of Birmingham has talked of the need for a learning government for a learning society because, he says, "we are beginning to recognise that we do not know all the answers". Permaculture is not a panacea, but it provides a discipline and structure that can help us to tackle problems a bite at a time.

Permaculture pre-dates Local Agenda 21, but it has influenced much of the methodology that has charactarised so much work within communities and local government since the Rio Summit in 1992. In 2000, the government introduced new duties and expectations on local government regarding Community Planning, Best Value, Quality of Life Indicators and Modernising Local Government. None of this has arrived with a blue print: all of it depends to some extent on experimentation. The design solution principles of Permaculture are as relevant to these new experiments as they ever have been 'to any personal or organisational review of quality of life.

This manual exemplifies the principle of more than one right way. There is similarity but distinctiveness; process but invention; and exploration for personal solutions.
In supporting the writing of this manual, WWF is hopefully demonstrating one of the things we do well: that is, supporting the dissemination of good, interesting and best practice from organisations that share a concern to increase the sustainability of contemporary lifestyles.

Ken Webster
Head of Local Government and Community Unit, WWF-UK
August 2000

Contents

Introduction

⇨ Introduction

This book is for existing teachers and budding beginners alike. It assumes that you have attended a Design Course and have a good understanding of the subjects covered, with practical experience in at least some of them. You probably don't have a formal teaching background, but this does not matter. The emphasis is on how existing teachers teach it. Very few of them have a formal teaching background, and yet they create practical, informative and inspiring courses.

This Guide contains a selection of tried and tested sessions developed by teachers in Britain. The authors aren't saying "This is how you should do it", but rather, "This is what works well for me". Adapt the sessions to suit your own style and the needs of your course participants.

○ Who can teach?

Permaculture is a copyright term, which can be used freely by anybody for non-commercial purposes. However, only graduates of a Permaculture Design Course can offer services for payment, such as teaching, and describe them as permaculture.

Graduates should also have two years of practical experience before taking on the role of lead teacher within a course, although they can take on an apprentice role during this time. It is therefore recommended that those wishing to teach find an experienced teacher who is willing to provide apprenticeship or training. Further guidance is given in Section 1.

○ The Design Course in context

Since 1982, when the first Permaculture Design Course was held in Britain, British permaculture has gradually developed its own unique style. Unlike Australia, where permaculture originated, Britain has a high density urban population, and gardens are typically small. This difference is reflected in the greater emphasis given here to people and community.

Always keen to minimise effort, permaculture designers draw upon the ideas, skills and experience of those around them, and in Britain we are fortunate to have a wealth of voluntary and charitable organisations which can provide graduates of a Design Course with a multitude of support. These are signposted during courses, and this makes functional the claim that we should co-operate rather than compete. Permaculture is part of the solution, not all of it.

Since the early 1970s a significant green movement has been developing out of an awareness of and concern about environmental destruction and social injustice. In 1992 there was official recognition of these problems when the Agenda 21 document was signed by the heads of 178 nations at the Rio Earth Summit. As a result of this official endorsement for sustainable development, many more people are questioning the unsustainable status quo, and looking more widely for solutions.

Permaculture designers are no longer seen as starry-eyed idealists, but as people with a useful contribution to make, and a fruitful collaboration is emerging between permaculture and many Local Agenda 21 (LA21) initiatives. For example Design Courses are increasingly being funded by LA21 officers as part of their overall strategy. Some LA21 officers are attending permaculture courses as part of their training. As permaculture projects and initiatives become more widely known, it is likely that this will accelerate the level of collaboration.

Permaculture education is also gaining wider recognition. The Open College Network has given the course accreditation (more details on pages 59-50,) and the Workers Educational Association is providing a route to run the course within Adult Education programmes. A number of universities are running the Design Course as a module within Architecture and Environmental Science courses. Bradford University and Richmond College being notable examples.

The Permaculture Association (Britain) will continue to promote the Design Course, and support the graduates, groups and projects which result from them. We hope that through collaborating with other organisations, such as WWF-UK, the Design Course can continue to inspire practical, common sense solutions.

○ Intentions of this book

· To collate existing teaching practice in the UK.
· To provide new teachers with a starting point for planning their sessions and designing courses.
· To provide experienced teachers with some new ideas to draw on.

The emphasis is on HOW we teach rather than WHAT we teach and has been written on the assumption that teachers already know much of the specific detail of the subjects. Where this is not the case, teachers have a responsibility to research the subject and familiarise themselves with practical examples and case studies, points of information to pass on and links to permaculture design principles and ethics.

This book makes this easier by providing references to recommended books and relevant organisations. The Association can also help in this process through the Teachers Group, which can set up training, link apprentices with teachers and act as a focal point for the collation of good practice and new materials. Many of the handouts mentioned in this book can also be obtained from the Teachers Group.

This is a guide to teaching permaculture, not an 'off the shelf' Design Course. It is important that each course is designed to be responsive to the particular needs of its participants. The core content – ethics, principles and design – will always be included, but the range of complementary topics and examples will change from course to course, and from teacher to teacher.

We have presented a range of approaches that different teachers take. For some sessions, we have deliberately included contributions from more than one teacher, to show different ways it may be taught.

There is no 'Permaculture Rule Book', and we have not attempted to standardise the terms different teachers may use, but have respected the richness of language used by contributors. For instance, there seems some overlap between what some teachers call 'patterns' and others call 'principles' – we have not tried to define where the line should be drawn between them.

We hope that this guide can make a useful contribution to the development of permaculture teaching in Britain. If you have any comments to make about this book, or about the continuing development of permaculture education, contact the Association, we would be very pleased to hear from you.

Andrew Goldring, June 2000

Aspects of teaching

Aspects of teaching

In this section teachers present some of the methods they use to develop the skills and understandings of the students, and look at some of the wider issues that need to be considered when designing a course.

'Do I Teach?' offers a slightly more philosophical approach, whilst 'What? Car Maintenance in a Permaculture Design Course!' provides some thought-provoking suggestions for future development, from a more analytical point of view.

The process of becoming a teacher is explained, with Association policy also made clear. Two examples of teaching permaculture with other organisations are presented.

Action Learning

Because it is so important to translate the theory and principles of permaculture into action, an approach that permaculture teachers and facilitators often take is one of Action Learning.

Tell me and I will forget
Show me and I will remember,
Involve me and I will understand
(Confucius)

Action Learning is a cycle which includes the following elements:
· Doing practical activities and project work.
· Systematically observing the effects.
· Thinking about how this experience affects our understanding of permaculture.
· Working out how these conclusions will affect plans for the next 'action opportunity' phase.
· Doing more practical activities that incorporate these new insights.

Graduates that go on to work towards a Diploma, follow an Action Learning Pathway. For more details see 'Working for the Diploma', pages 366-367.

The First Day

Graham Bell

The first day is very important within the course for a number of reasons. Day 1 sets the tone for the experience – and that is what a good course should be, a profound experience, not just an extended learning session. You can set standards of behaviour which will enhance the process for participants, students, teachers and supporters. You can establish models for good learning. You can learn as a group about people's experience, ambitions, expectations and capabilities. Good training is participative – start as you mean to go on. When the task is shared the burden is lighter.

You cannot do these things well, however, without good planning and preparation.

A question to consider

Is the Design Course training or education? I define training as the process of instructing someone to be able to do something, and education as the acquisition and understanding of information. I believe that the more this course becomes training, the more successful it will be. Successful for whom? Ultimately the person who matters most – the individual student. The measure of success is then "How well can each student do the job we are talking about?" This measurement can be offered up against any one activity, eg "Interpret contours on a map", or the course as a whole, eg "Make effective permaculture designs."

I expect most teachers find, like me, that they have a vocabulary of techniques and tools for achieving their instructional objectives – oh, and that's the first one – to have instructional objectives. I expect also that you will find this 'toolbox' is constantly evolving, as you improve your training skills through experience and observation.

In this section I will suggest some underlying principles, and describe some techniques I have used to fit them. It's crucial that the methods you use 'fit' you, so that you can deliver programmes comfortably and well. We all need times when we can try out new ideas and methods to achieve that degree of fit in future. The first day may not be the best time for radical experiments, although sometimes that will be unavoidable! Principles are in constant evolution. This section concludes with a list of exercises and where to find them in the handbook.

There are three groups of principles:
· Things we decide are essential to do.
· Things we decide are useful to do.
· Things we decide are best to avoid.

Essentials

· Make sure everyone on the course knows everyone else's name. Have some good name games to play. Starting with play sets a light tone and contradicts people's essential fear of education derived from harmful prior experiences.

· Inform participants of domestic arrangements. Avoid taking responsibility for these – it makes it difficult to focus on teaching well. Do state the obvious: eg "If you want the toilet please just go, there's no need to ask. We're not in school now!" Invite participants to accept responsibility for the little jobs that need doing, which if shared are easy.

- Give essential health and safety information. As instructor you have legal responsibilities for the health and safety of participants. Point out first aid facilities, and establish ground rules on any no-go areas, access to tools, (no) smoking policies. I take it as read that recreational drugs, including alcohol, and good learning are not contemporary activities. In some circumstances it is necessary to state it.

- Agree a timetable for the duration of the course. Make the territory safe for the participants. To do so it is also necessary to be sensitive to what that means to different people. If you want people to be honest and open about their fears and aspirations, their successes and failures, their personal experience and so on, then it has to be a safe environment for them to do that.

- I point out that we are all going to work very hard, and the hardest thing we are going to do is listen. I thank people in advance for the effort they are going to put into that. I invite people to hear really well what each person is saying and to avoid the temptation to agree or disagree with points of view, rather to give even respect to each voice in the room (or out of it). I suggest that if information is given which people feel is not useful to them personally then it's perfectly OK to leave it to one side and concentrate on the things which are more useful. Most importantly it is not essential for anyone to be 'right' or 'wrong' and the words 'should', 'must' and 'ought' will be avoided for the duration of the course. This course is about choices and how we can be as well informed and skilled as possible to make them.

- Your timetable will not work if it doesn't suit people, so make some time early on to review a draft copy.

- Establish methods of training which will be used. Make sure that on day one your styles and techniques are made clear and accepted.

- Get everyone participating.

- Learn why everyone is there. It's rarely as simple as "To find out what permaculture is" (they could read a book). I find permaculture Design Courses attract people with three main aims:
  - To enhance their work skills.
  - To develop their own plot or homestead.
  - To give meaning to times of pivotal change in their lives.

- Generate the atmosphere which will make effective learning happen, which is different from making everyone happy or comfortable. All the above points are part of this. One technique I use is to generate a discussion about 'what constitutes a good course' and then to use that feedback to explain pattern languages. I call this pattern 'The Flywheel Effect'. (See pages 176-177.) The overt motivation is to get people to define what does and doesn't work well, and therefore to get their commitment to supporting a well-designed process. A spin-off is to get good feedback on what their expectations are.

- Exercises which encourage core skills: listening, observation, making links. Permaculture is not about knowing encyclopaedic lists of species or how to construct a solar chimney, though these skills all help realise a design. It is essentially about skilful observation, and being able to make the best of what is available; to think ahead and vision change, growth and decay; and to fit these to the needs of real people. Emphasise these skills from the outset by offering tasks and games which emphasise these abilities.

Useful

- Encourage even participation. If everyone participates, everyone is encouraged.

- Find talents in each person that can be used to make them co-teachers. The collective wisdom and skill of the group always far outweighs the teacher.

- Discover design opportunities on site, indoors and out. It's much better to work with reality you can see than designing fictional or remote examples.

- Establish shared responsibilities. Shared work builds group dynamics, and saves someone else having to skivvy.

- Strike a balance between being structured and flexible.

- Find early opportunities to build self-confidence in all students. Ability needs to be supported by self-confidence if students are to be able to 'do' permaculture after the course.

- Establish that ultimately you are in charge. This is not an ego thing, it's more about achieving aims safely.

- If any deficiencies in the venue are identified, try to resolve them early. Get the group's co-operation in making improvements – always a good practical exercise in working with what you have.

Avoid

- Being seen as the person with all the answers.

- Making assumptions about people. For example the noisy person may actually be lacking in confidence and the quiet person just happy to listen. It's easy to assume the opposite. Try to remain open. If anyone 'behaves badly' remember it's the behaviour that is undesirable not the person.

- Favouring individuals. Similarly notice if anyone is 'an especially good student' in your mind, and be careful not to overuse their input.

- Allowing destructive criticism of the host(s) and their site. Don't let the hosts be made to feel that in some way they're doing it wrong or inadequately. Ask permission before interrogating hosts as 'clients' or in making critical comments.

- Put downs, from yourself or others. Try to find positive ways to express views on any inadequacy. Encourage others to do the same.

- Arguments in teaching time. If something is going to run and run, suggest it as a topic for tea break or evening discussion.

- Measuring success by how much information you give. The real success indicator is how well can the students do what you are trying to teach.

- Language which offends. Offensive language in one place is quite acceptable somewhere else. It's really a question of being sensitive and knowing when to back off. Sometimes you have to go in and 'lance the wound' or the whole show grinds to a halt. It's better not to get to that point if you can avoid it.

- Judging yourself too harshly. Many people have at the end of a permaculture course produced design work which was, to the teacher's eye, disappointing. Remember the course is just the start.

What happens after may be wonderful. Don't put yourself down because of perceived failures and mistakes. Remember mistakes are really important. Without them there is no progress.

○ General

- If possible start a course with an evening meal and slide show, finishing quite early. People are dying to know about who's doing what, where, when and how successfully. They also want to grasp the whole immediately. Pictures help you do this. If this facility isn't available give a short talk about how you were inspired by permaculture, and your personal experience of it working well or challenging you.

- Always tell people to be at the start on the first morning half an hour before you need to get going, but don't tell them you've built in a 'buffer zone'. Stick to session start times thereafter, ie encourage punctuality in practice. On Day 1 start only when all are present if you possibly can.

- Don't feel you always have to be available, or that you are responsible for each person's personal issues.

The important principle I take from Lao Tzu, the Tao master, is "You know a good leader, because when they have gone, the people say they did it themselves".

○ Day 1 checklist

- ☐ Have you got all the materials you need?
- ☐ First aid?
- ☐ House rules?
- ☐ Outdoor exercises?
- ☐ What's your timetable for today?
- ☐ What might go wrong? Strategies for coping?
- ☐ Who supports you?

- ☐ Handouts?
- ☐ Toilet facilities?
- ☐ Name games?
- ☐ What do you want to achieve?
- ☐ Happy with the teaching space?
- ☐ What else can you do to make it go well?

○ 'Toolbox'

This is not exhaustive, but mine contains:

- ### Attention span
  The amount of time for which you can concentrate on any material. It varies between each of us, and for each of us will change depending on our level of interest in the material, how tired we are, whether we find the presentation method stimulating and so on. During the day we try to ensure there is a mixture of activities. Avoid places with too many distractions. Movement, change of scenery, fresh air and exercise and also some stationery listening, speaking or contemplating activity. Games, dances and songs are useful in breaking up sessions, and generating fresh energy. Be aware that not everyone is equally happy with techniques with which they are unfamiliar.

- ### Checklist
  A safe environment.
  A good environment for learning.
  Adequately resourced and supported teacher(s).
  A timetable which meets the needs of all.
  Participants with adequately common purpose.
  Good two-way communication between all parties concerned.
  Materials required for learning.

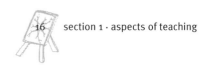
Domestic arrangements to suit the participants.
The course objectives are met.

· **Coffee time**

This is a generic term for mid-morning break.

· **Crayons**

Getting people to draw in a childlike way often unlocks creativity. It's very important in getting people to express their dreams, as in My Dream Home. (See page 278) This helps people discover "What do they really want?".

· **Flipchart/Black or white board**

These may just be ends of newsprint reels, or something more hi-tech. Useful to have something light and portable for mobile sessions.

· **Games**

It's good to have a bagful to choose from. They help attention span, build group dynamics and can make

valid points or help in other ways, eg trust building, conflict resolution. (See below and page 371 for references to books on games and ice-breakers.)

· **Goals**

It's helpful to get folks to leave a course with a set of written goals. Allow at least 30 minutes for such a BIG affair, including feedback. Structuring can be very helpful.

· **Handouts**

Some love 'em, some don't. Make 'em simple and you'll get told off. Make 'em complex and you'll get told off. They are ultimately worthwhile to reinforce what you are talking about, or to give detail which doesn't fit into lesson times. They can take a long time to produce, if done properly, but are worth building up over time. They can also be expensive.

· **Health and safety**

As an instructor you are legally responsible for the people you are teaching whilst they are with you, and for what you teach them. This doesn't mean that their problems have become yours for now and evermore. It does mean you must act safely and healthily and legally, encourage them to act safely and healthily and legally and not teach things which are unsafe, unhealthy or constitute illegal activity.

Do not be afraid to state the obvious when you enter issues in this area, eg if demonstrating pruning explain how to use and sharpen secateurs, and the importance of keeping them locked shut when not in use. Just because you are a dab hand at it don't assume your students are. Being very basic about these issues sometimes sounds daft to you. It won't to your students.

You should have public liability insurance, and so should the responsible person(s) on whose site you are.

You should have training in emergency first aid, principally resuscitation, and have a basic first aid kit to hand. Do not give first aid to anyone unless you are a properly trained first aider. Always establish who such people are in the first session of the course. If not available, issue instructions for the nearest available emergency facility.

If indoors point out any fire exits, evacuation procedures and assembly points.

If the course includes children there should be adequate adult supervision at all times when the

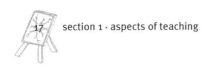

parents are occupied. It should be absolutely clear, preferably in writing, when the parents resume or relinquish responsibility for their children.

· Hum group

Folk are divided into pairs, and one group of three if you have an odd number of people. They are given a topic to discuss for five minutes and invited to give between four and six points relative to that topic. Help is given only if required. When the 'hum' subsides the contributions are taken by each pair in turn giving one point, going round a couple of times, to ensure each person has spoken, then opening it to the floor until all points have been gathered onto a flipchart. The instructor asks contributors to clarify their offerings. This is a good introductory technique.

· Introductions

It's worth telling people your name, your experience and how much you value their participation.

· Objectives

We have a number of objectives on a Design Course. Some of them we may all share. Others may be particular to us as individuals. Being clear about objectives is important to a successful course. How will you know when you get there if you don't know where you're going?

· Open questions

There are six honest serving men and true.
They taught me all I knew.
They were HOW WHY WHAT WHERE WHEN WHO.
Open questions are important because they require the respondent to give information, rather than Yes or No which are answers with minimal progress.

· Opening questions

It's good to have a stock of questions that get people talking about themselves and what matters to them.

· Overhead projectors

They are expensive, but useful. Although difficult to use in the field, they are however excellent for presenting material which is used over and over again.

· Participation

Students who participate must be alive. Retention is better amongst those who participate. Participation lets the knowledge of the people in the group do the teaching for you as much as possible.

· Permission

These kind of permissions: "It's OK to make mistakes", "If you want to go to the toilet just go, don't ask" are very important in setting people at ease. Many people have bad residual memories of educational processes and will welcome a bit of encouragement and adult informality. Give people permission as often as possible, eg "I'd like to invite you to do XYZ..." rather than just "Do it!" That's your fall-back position.

· Planned visit

They can be structured and therefore made more relevant by giving specific points to look for before departing, and reviewing after the trip or on the site. This is to encourage students to evaluate what they have seen and assess possibilities for improvement. This is a form of practice which can help give reality to design exercises.

· **Preparation**

There are a number of predictable things which can go wrong. Good Day 1 strategies seek to minimise the risk of these, and build in solutions. No amount of preparation can cope with everything. However, in general terms, more is better than less. Estimate a ratio of four times preparation to session time when writing new material.

· **Songs**

They aid attention span, may be fun and should be used to encourage all to 'find their voice'. They can also teach and help group dynamics. Invite group contributions.

· **Structure**

Structured sessions are helpful in achieving specific objectives, and may make the material better focused on particular learning outcomes. Free flow work tends to be more effective where a 'taster' is being offered, ie more educational based than training.

· **Tea time**

This is a generic term for afternoon break.

· **Tick box**

Seemingly opposing statements are offered for students to favour. Their answer is not important in terms of which they tick. It is simply a tool to promote debate. Useful in uncovering belief conflicts and value issues for debate.

· **Homework**

By which I mean effectively offering a theme for evening consideration and meditation. I use it as part of teaching zonation which I consider a very difficult subject to teach well. So I start on the first evening by inviting people to consider themselves and thereafter we work outwards into the universe in stages.

An example of a homework session:

· Session: 'Me, My Greatest Asset' – recorded meditation.
  Objective: Students will be able to name ways in which they are their own greatest resource. Consider yourself. Think about what you own – but not just property! What about your skills, relationships, knowledge, tools, personality? Record what comes up as creatively as you like. If you find it difficult to unlock the answer try writing with your non-dominant hand or drawing with crayons. These techniques often effectively release our sub-conscious thoughts.

· Session: 'Me, My Greatest Asset' – review. Group presentation.
  Objective: Students will understand the value of, and be able to use dynamic means of, expression for passing information. They will feel comfortable with supporting others in this role. In groups of between two and four present in one minute per person a celebration of each person's value.

○ Potential exercises for Day 1

Please note that these examples are for selection purposes. It is very important not to attempt too much on Day 1.

Do include:

· An overview on what is permaculture, encouraging individual contributions.
· An outdoor observation exercise.

- A listening exercise.
- Statements of personal experience and goals.

The exercises are scene setting. They get people talking. They allow me to monitor who needs more encouragement, who needs holding back to allow others space; whether there are any big issues out there, or potential conflicts over widely differing mores etc. For instance we always plan space for The Vegan Debate.

○ Examples of Graham's sessions can be found at:

○ Further research

Let's Play Together, Masheder, M, Green Print, 1989
Playful Self Discovery, Platts, DE, Findhorn Press, 1996
Songs for the Child & the Child Within (Tape), Stuart, L, Findhorn Press

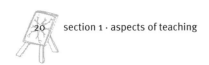

Notes from the Front Line – Permaculture Design Courses in the Inner City

Angus Soutar

As far as Least Change For The Greatest Effect is concerned, I teach in cities. And working in the inner city makes me confront problems I would otherwise fail to acknowledge. It is also great fun, because cities tend to be packed with interesting people.

I have found that adapting existing practice to inner city situations has stimulated more thought and consumed more of my energy than the transition from say, rural to suburban environments. This is all going on in parallel with an expanding interest in sustainability in general, where I am seeing increasingly diverse groups of people become interested in permaculture. This diversity of cultural backgrounds is a challenge to some of the more formulaic approaches that I have adopted in courses in the past.

In the interests of 'energy-efficiency', I offer these initial notes, based on my own experiences and those of my friends. I would be glad to share experience with others in developing them further.

Dealing with the expectations of students

No matter how clear the course promotion material is, I still meet people on introductory courses in the inner city who think that permaculture is all about gardening. Meeting their expectations can be a serious problem on an inner city course. By incorporating the various techniques of listening, thinking and talking mentioned in this manual, you can pick this up in the early go-rounds. Also inform students early in the course about the types of design exercises that will be done. Gardening is part of the course, but probably a smaller one than in other settings. In most cases, individuals can be accommodated, but they may become disruptive. "When are we going to get to the gardening?".

The above can also be true of people who have 'one track minds' about other issues in sustainability.

Course structure

The course in an inner city can be basically the same one as runs in a rural environment. Instead of incorporating a module on 'rural permaculture' for the urbanites, I like the idea of incorporating a weekend away at one of the rural sites listed in 'Permaculture Teaching Venues' (pages 368-369). Interesting site visits can also be programmed to bring in some variety. In London for example, we once visited a cemetery: it was the nearest thing to a wilderness for miles and miles. There are some good social structures to visit in cities, and I get a lot of learning myself on those visits. I see teaching organisational and village design as central to urban courses, but other teachers may decide there is little of that belonging in the core syllabus.

When planning the course, Peter Harper's comments in 'What? Car Maintenance in a Permaculture Design Course!' (page 36) are particularly appropriate – read them now and follow them up.

Observation exercises

Visit scrap land, derelict sites. If you get there before men with strimmers come to make things 'neat and tidy', these sites can be a major source of inspiration for nature's ability to regenerate even the most hostile areas. There are also plenty of examples of poor design: "Why did they do that"?

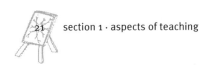

○ **Design exercises**

Land access is often a major problem in inner cities. There's always a bit of land somewhere and the students can have fun designing there. But will any of the designs be implementable? If sites are going to be productive, will we have to spend a major portion of the budget on security fencing and other measures? And if you're thinking glass, do you need to think polycarbonate? Or else a much higher fence? You end up thinking expensive.

But as we know, the problem is the solution – find exercises other than land design. As Bryn Thomas suggests in 'Zone 0: Retrofit Exercise' (pages 284-286), retrofitting buildings is useful and the design of social and economic systems is always possible (eg 'Community Economics', pages 329-332). If there is someone else in the group who knows more than you do about some technical aspects, that's helpful. I let them be the technical expert for the group and I concentrate on the permaculture.

○ **People power**

People are the biggest resource in any inner city area. As the Agenda 21 framing documents point out, people need the capacity to look after their own environment. Often, this capacity is more than just exercising technical skill. It's about people organising themselves and gaining social and political credibility. Energy put into helping people create their own beneficial linkages can be highly productive. What students do on the design course may not be as important as how they go about it, and what they learn from that.

○ **General approach to the course**

Beyond the basics, I like to be sensitive to the context of the students – are they 'activists', 'professionals' or 'disadvantaged' groups? The course shouldn't have to change much. The key point is, if the delivery team is sensitive to cultural background and language, then this can help with the participant's learning.

The design course is only the beginning of the process, I try to avoid overloading people to start with. Leave them questioning and ready for more, but later.

As for research to investigate the suitability of the above proposals for inner city courses, my recommendation is to try things out and see what happens – there's nothing like a little bit of 'vuja de'.

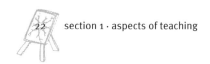

How I Teach

Patrick Whitefield

Aim

It's not possible to teach people everything they would like to know about permaculture in 72 hours. What is possible is to give them a new perspective, and a skeleton which they can flesh out with specific practical skills.

The perspective is the unique contribution of permaculture. It's not obtainable anywhere other than on a design course, whereas the practical skills are widely taught elsewhere.

The practical skills, such as gardening, forestry, landscape design and so on, can be ones which people already have, or ones they can go on to acquire later. I find it quite straightforward to teach a course which is equally appreciated by people who practice relevant professions and want to put a new perspective on their skills, and by complete beginners.

So my aim in a design course is to give people four things:
· The principles.
· Enough examples to make the principles come alive.
· Access to more detailed information.
· Hopefully a measure of inspiration.

Teaching style

There are many different styles of teaching. At one end of the spectrum there is the information-led style, which involves a lot of talking on the part of the teacher and not much else. At the other end is the process-led style, in which the teacher rarely – or in extreme cases never – imparts information directly to the students.

Both these extremes are to be found in the permaculture world. The first is motivated by the belief that there is so much valuable information to get across, and so little time to do it on a 72-hour course; the second by the belief that people don't actually learn very much by simply being talked to. There is truth in both of these beliefs.

My approach is to take a middle course. People do like to go away with some information at the end of the course, and talking is an effective way of getting it across – as long as it's done in moderation. Beyond a certain point people don't learn more by being told more, and I make straight talks less than 30% of the total teaching time.

People also need to participate in their own learning process, and simply listening and taking notes is a pretty passive mode of learning. A variety of teaching methods not only makes for a lively and enjoyable course, it's also the most effective way of teaching. If people go away feeling they've learnt it themselves they're more likely to use it.

Methods

I base my teaching on the idea that we learn in different ways:
· Verbally, by listening or reading.
· By relating the idea to our own experience.

- Visually, by seeing examples.
- Practically, by doing it.

All of us learn in all these ways, but each of us is to a greater or lesser degree a specialist in one or two of them. I aim to teach the major subjects on the course by a mixture of methods which are suited to these learning styles:

- Verbal – talks. I draw a mind-map on the board as I go, plus diagrams and simple illustrations where appropriate.

- Own experience – discussions. After talking for a while on a subject, I ask the students to get into groups of three and discuss how they can put these ideas into practice in their own lives. I have found this a very powerful learning method.

- Visual – slides, or actual examples on the ground. I have slide sets of varying length on most topics in the course. Some rate sessions on their own, others are used to support talks and discussions. The importance of teaching on a site where permaculture is practised, or organising good visits, can't be overestimated.

- Practical – exercises. These are mainly the design exercises, but also I am starting to build up a collection of small desk-top exercises.

Other subjects may be covered by one method only, eg discussion or exercise, or two, eg talk and discussion. Methods I use in addition to the four above include:

- Discussions:
  - In pairs. People find this pretty intense, and it needs to be used carefully.
  - In threes. I often get them discussing in threes, on a variety of subjects, not just "How can I put this into practice". (See above.) Groups often evolve to anything from two to five, which is OK. But above five I split them up.
  - Discuss and report-back. I split them into three or four groups. I ask each group to discuss a particular subject and then give a short presentation to the whole group. Points to watch are:
    a) it takes longer to cover the same amount of information by this method than by a talk followed by discussion in threes.
    b) for some subjects the students need previous knowledge.
    c) it doesn't work if the subject is too wide – eg "Make a list of the inputs, outputs and characteristics of sewage" works well, while "What has permaculture to say about the design of new woodlands?" is too wide.
  - Whole group. This needs to be used with caution, as a) it can be dominated by one or two people, or b) some students may feel it's getting vague and irrelevant.
  - Read and discuss. For two topics on the course I get them to read a handout there and then, and when they've finished reading discuss whatever arises in the whole group. This usually works very well. (See 'Biodiversity' on pages 201-203 for an example.)
  - Spontaneous. Sometimes a talk will turn into a discussion. This is great, but if it gets off the original subject of the session it may mean that some of the intended course content is lost. When this starts happening I ask the group as a whole whether they want to go with it or return to the

subject. There may be just one or two articulate people who are interested in the discussion, and others may later quietly resent having missed something they want. You can only find out by asking.

· Brainstorms. These often come at the beginning of a talk. I ask people to give me all the factors on a particular subject, eg "What makes good listening?". It's important to make a clear distinction between a brainstorm and a general discussion. In a brainstorm I keep to the following rules:

a) I write down everything the students call out, even things I disagree with.

b) I don't comment on or discuss anything during the brainstorm – that comes later.

c) I use the same word or words that the person speaking used – if I need to condense it I ask them if my wording is OK before writing it down.

d) if I want to add anything of my own I do it towards the end.

e) I may call a halt when I feel we've got enough down, as long as no-one's bursting to add something, and everything which has been said already has been written down.

A brainstorm is a space for students to express themselves freely, and any editing of what they say, or any drift into discussion before they've had a chance to say it, is quite rightly resented.

· Handouts (available via the Teachers Group, see pages 48, 49 and 364)

  · General handouts. 14 sheets. These are ones which are handed out to every student. They come in three kinds:

    a) information to supplement that given in talks, eg 'Energy, Definitions and Figures'.

    b) material for read and discuss sessions (see above).

    c) reference – Booklist, Useful Organisations, Suppliers of Plants and Seeds.

    Most of these are written by me and I am happy to share them with other teachers.

  · Optional handouts. Some 80 titles, mostly photocopies from various sources, with more detailed information on specific subjects. Students can look at these and select what they want. The handouts are copied during the time of the course, at a small charge to cover costs and time, usually by a student who wants to earn a little money.

· Activities. In addition to the exercises mentioned above, a very popular activity is a session of practical work, ie getting your hands dirty on a communal task. Whenever I include it on a course some people always say it was the high point of the course for them, and if I leave it out there are usually some who say they would have liked it.

· Videos. I use videos in one of three ways:

  · to reinforce material taught by other means, eg the Coppicing episode from Spirit of the Trees.

  · to cover subjects I don't teach, eg Global Gardener for tropical permaculture.

  · to provoke discussion, eg Ancient Futures. (Use this one with caution – people can get depressed.) (See page 371 for full details of these videos.)

 Course design

(See my 'Example Course Format' on page 59.)

There are two main themes to the course:

· Information input. These sessions are mainly in the mornings. They start with ethics and principles and then go progressively through the zones so that all relevant material is covered by the beginning of the Main Design Exercise.

- Practical exercises. These are mainly in the afternoons. They start with learning the basic skills of surveying, observation, listening etc and culminate in the Main Design Exercise.

Each day is planned so as to:
- Have a main theme, eg gardening or woodland, though there will be some sessions on other topics.
- Have a mix of teaching methods.
- Take account of the changing energies through the day.

An example of how I do this in practice is Day 2 from my Example Course Format. The twin themes are Principles and Gardening. These go well together, as most students are familiar with gardening, and find it easy to see the principles in action in that context.

9.30 Principles Part 2

The first session on any day is the best opportunity for solid information input – though you can't give them that every single morning. I teach the principles by a mixture of talk, slides and discussion, which makes this session much less solid. But the amount of information involved means it's still pretty dense.

11.00 Morning Break

11.30 Gardening Part 1

Gardening 1 looks at the key planning tools – zone, sector, network and elevation analyses – as they apply to gardening. It's about two-thirds talk with sketches and one third discussion. The talk is immediately after the morning break, and was preceded by slides and discussion at the end of the principles session.

What Do I Eat?

This is a desk-top exercise followed by discussion. It's placed here because:
a) it helps people to relate some of the principles to their own experience;
b) coming early in the course it gives them something of a personal starting point;
c) it provides an activity, in contrast to the intensive information input of the morning so far.

1.00 Lunch

2.00 Surveying Skills

The first session in the afternoon is almost always an outdoor activity. This is the time when people are most likely to nod off, and the fresh air and movement counteract this. It's also time for a change. In Surveying Skills students learn things they will use later on in the course, particularly in the Base Map Exercise.

3.30 Afternoon Break

4.00 Slide show and video

The second half of the afternoon often consists of a slide show followed by a video. People are less fresh than they are in the morning, and visual media are easier to digest than straight talks, and require less effort on their part than discussions.

An Urban Garden

This is a slide show of Michael and Julia Guerra's garden, an inspiring example which many people can follow in practice. It illustrates much of the material covered in the morning.

Global Gardener Part 1

Dealing with the Tropics, this video gives a change of scene and adds some variety to the day.

I don't normally include evening sessions. The course is intensive enough without that. Six hours' learning is about as much as most people can do in a day. If you go much over that they don't actually learn any more.

Timing

I think almost every new teacher of permaculture is surprised at how long it takes to put across a certain amount of information. You think you haven't got enough material for the session and find you have enough for two. In fact it's a very good idea to have more than enough material, but the result is often that sessions over-run. I also find that spontaneous discussions have a habit of cropping up just when a session is supposed to end thus causing over-runs.

With experience I've got a better idea of how to match input with the time available, and I plan the discussions in, but over-runs still happen. And so they should! I'd rather let myself chose whether or not to go with what's happening than be confined in too rigid a timetable.

To cope with this I have a few 'Spare' sessions in the course. If things over-run to the extent that a whole session has to be postponed, I move it to one of these. In the rare event that nothing over-runs there are plenty of extra topics which can go into a Spare session.

Sometimes over-runs can leave 15-20 minutes to fill. Rather than doing half the session that's on the timetable and leaving the rest till later, it can be better to postpone it and slot in a short session which is kept in reserve for a time like this. For example, I have a session on Leaf Curd which I rarely put in the timetable but always include at some point during the course.

Some sessions always need more than 45 minutes and others less. I often put one of each next to each other in the timetable, knowing that they won't each take exactly one session, but that together they will fit into the hour and a half allotted.

Content

My priority is not so much to give people answers, as to help them to ask the right questions. Asking the right questions is absolutely fundamental. If we ask the wrong questions we can't hope to come up with viable answers.

To a great extent the detailed information we need in order to come up with answers is out there, and it becomes more accessible with a good booklist and contact list of organisations. But an overall holistic perspective is much less easy to acquire, and a Design Course is one of the few places where it's available.

Most of the material I include in my information sessions is either fundamental concepts or a series of questions to ask about a subject. See for example my session on 'Energy' (pages 271-273).

Sometimes people have asked me "Are you telling us we should do such-and-such?" I always reply that I don't want to tell them what to do, but to help them become better equipped to decide for themselves.

Nevertheless, on some courses there is a demand for more facts, and if the demand is general enough I bow to it and somewhat alter the sessions I have planned. If I do give more detailed facts it's mostly on subjects which are permaculture specialities, such as micro-climates and windbreak design; information which is less readily available elsewhere.

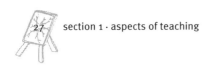

○ Feedback

Mid-course

Course students rarely tell you they want things done differently if you don't ask them, so a mid-course feedback session is useful. Minor changes to the course can make a big difference to people's enjoyment of it. Even if some of the proposed changes are not feasible, people appreciate the opportunity to be heard and get an explanation.

On the other hand it's good to be cautious about making major changes during the middle of a course, especially for a relatively inexperienced teacher. Even if the suggestion is a good one, the plan you carefully worked out when preparing the course is more likely to work well than a modification made quickly amid the stresses and distractions of the course itself.

It can be easy to be swayed by one or two students who have particularly strong views about what you ought to be doing. On one of my courses there was a student who was constantly on at me to teach more sessions by the discuss and report-back method rather than as talks. In the end I gave in and changed a planned talk to a discuss and report-back. But the subject wasn't suitable for that method and the session was a failure. I'd made a change for the wrong reason – not because I felt it was the best thing to do, but because of pressure from someone else's agenda.

End-of-course

Feedback is essential if the quality of courses is to go on improving. I usually use a combined written and oral feedback session. First I ask the students to write down their comments about the course content, teaching, venue, administration, domestic arrangements or any other subject, under the following headings:
· What I particularly liked.
· What I particularly disliked.
· Anything I would have done differently, eg anything missing.

I write the three headings up on the board. Some people take longer than others, so when most have finished I ask those still writing if they would like to finish later. I then ask if anyone has anything they would like to say directly to me and the other course staff. This is voluntary, but most people usually say something.

Back at home I collate the feedback. For each design course I note:
· How many students were on the course.
· What comments were made under each of the three headings above.
· How many students made each comment.

Thus a comment made by half of the students on the course usually carries more weight than one made by a single student – though not necessarily.

I keep the notes from all design courses together on one document, so any consistent patterns can be easily seen. If I see the same comment coming up on more than one course, I know I need to pay attention.

○ Changes

Changes to the way I teach don't only happen in response to student feedback. Some, such as introducing desk-top exercises, have been deliberate decisions. Others have started out as a response to the needs of a particular course and gone on to become regular features. Discussions in threes, for example, arose when I was preparing for an evening class in winter and had nothing in my repertoire except talking and slide shows. It worked well and I've used it ever since.

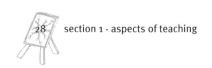

Though I'm cautious about making changes at the mid-course stage, I do make them if I'm sufficiently confident they will work, and they can also become part of what I do regularly.

I did this on a recent Introductory Course. The group was well-informed and articulate, and the weather had kept us inside more than usual on Saturday, so I felt Sunday morning needed lightening up. The first two sessions were planned as talks with discussion in threes on Gardening and Community Agriculture. I changed them to a discuss and report-back on 'Gardening' and the 'What Do I Eat?' desk-top exercise. The subject matter for the gardening session was how the concepts of zone, sector and elevation can be used in the garden, so there were ready-made topics for three discussion groups. The morning went well, and didn't over-run more than usual.

I don't know yet whether this will become a permanent change in the way I teach Introductory Courses, or whether I will now do the 'Gardening' Part 1 session on Design Courses as a discuss and report-back. It will probably depend on the nature of the group. But I do know it works, and I have another option open to me.

Bibliography

1 Spirit of Trees
2 Global Gardener
3 Ancient Futures
Please note that full details for frequently referred to books and videos can be found on pages 370-372.

Teaching a Permaculture Design Course

Joanne Tippett

On the courses I teach, the emphasis is on the process of design. The educational approach focuses on the quality of holistic thinking, through the application of ecological principles and the understanding of patterns.

Course structure

This course structure was initially developed in South Africa and Lesotho. It has been used to teach students from a wide range of educational backgrounds and literacy levels. The course is tailored to the interests and needs of the students through the examples used to elucidate principles and in the content of the design exercises.

Principles of permaculture form the backbone of the first week of the course (see page 58). The Natural Step model of sustainability is used to elucidate permaculture ethics (see pages 89-90, and Ref 1). Teaching about patterns provides the foundation for understanding permaculture principles and the design process (see pages 222-224).

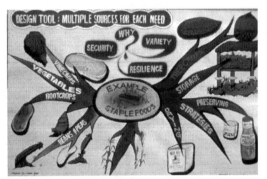

Sessions on principles are interspersed with teaching about ecology and the ecosphere (water, soils and the atmosphere), and lessons for design which can be derived from this understanding. The technique of mind-mapping, developed by Tony Buzan (Ref 2), is taught at the beginning of the course. A mind-map is a pattern-based graphic technique for representing ideas and information, which mimics the way the brain stores and integrates knowledge. Mind-maps are used in many of the educational materials and exercises.

The second week is structured around an extended practical exercise in the process of permaculture design. This practical is based on the methodology I have called SuNstainable Pathways Design'.

'Pathways' integrates goal formation, contextual analysis, participatory design and permaculture design tools in an easy-to-follow, step-by-step process. The process takes the form of a pattern language, which uses templates to collect and analyse information and generate design possibilities.

These templates integrate mind maps and ecological principles. The use of large templates, with movable pieces, facilitates communication between people in a group and between different projects and groups. The process encourages developing connections and relationships and laying out parallel, or alternative, possibilities which are tested against the previously formed goals.

Teaching techniques

The various techniques are designed to actively engage students in the process of learning through individual and group exercises. As the course progresses, posters of principles and mind maps created during group sessions are hung on the wall. By the end of the course, the classroom is a kaleidoscope of colour and ideas. Drawings and maps uncover the trails of innovation travelled. From this map of the learning process, principles can be reviewed in relationship to each other and to the practical exercises on the course.

Complex information and concepts are broken down into simple parts, or layers, which are taught in sequence to build up a coherent picture of the whole.

Most sessions start with a group exercise, which leads into a principle. These exercises are designed to help people understand the process of design by teaching principles on a physical, as well as an intellectual, level. This includes building up a large graphic representation, or model, of the principle. Examples of these exercises can be found in my section 'Principles of Permaculture'. (See pages 85-90.)

Each exercise leads into a class discussion, with the facilitator emphasising patterns and emerging themes. The degree to which the theoretical underpinning of the principles and design process is taught varies with the interest level of the group. Basic ideas, eg looking at systems as wholes rather than as collections of independent elements, are taught both directly and through exercises.

Applications of principles are shown through case studies, slides and practical examples. Sessions end with a discussion of links to previous principles. It is important to leave time for students' questions and ideas.

Uncovering people's knowledge

Teaching techniques employed on this course aim to uncover what people already know about ecology, showing how this knowledge can be made more useful in design through the application of permaculture principles. This process includes emphasising connections within and between participant's different areas of knowledge and permaculture. Information about students and their interests gathered at the beginning of the course is invaluable in this process.

One of the ways in which knowledge can be made more available for use is through the teaching of thinking tools, such as those developed by Edward de Bono (Ref 3). These tools encourage creative design and lateral thinking, which complement more traditional analysis.

Part of this uncovering process includes asking students to describe various aspects of their experience, with the intention of making underlying assumptions and perceptual filters visible, so they can be examined and compared to holistic, ecologically based paradigms. This 'making visible' can be done in a literal way, by asking people to write down their thoughts and ideas on a particular topic onto strips of paper, then making group mind-maps from these strips, using colours to highlight paradigms and assumptions.

Building a framework

After teaching a principle, eg edge effect, or concept, such as holism and feedback loops, the mind-maps of the group's thoughts (described above) can be revisited. Alternative views can be demonstrated, reexamining assumptions through the filter of permaculture principles and systems concepts. The use of strips of paper and blu-tack means that it is easy to tear apart groupings and ideas and to rebuild them in different ways, adding connections and new ideas. This method encourages students to question their knowledge base and learn how to make connections for themselves.

The process of making flexible mind-maps, then returning to them for adjustment, sets the scene for the design process, in which many of the early steps are revisited as the design unfolds, allowing for changes in earlier stages and checking new ideas against previous steps of the design process.

Feedback and cycles of learning

An important component of this learning process is the building of feedback loops into the structure of the course. This leads students through a process of understanding how each principle fits in with the rest. This has multiple functions. The repetition of principles, combined with showing different aspects of their application, reinforces learning; and a 'thick' knowledge of relationships and patterns develops from the interaction of simple parts.

An example of such a loop is: showing how the resilience in a system which comes from designing multiple sources for each need is linked to the need to design for surge and pulse, or extremes, eg flood and drought. This is later linked to the idea of developing a local and biological resource base, which provides for greater stability in terms of long-term resource access. This is tied together with a discussion of genetic diversity and its links to cultural diversity and the way in which the culture of an area is tied in with its resource base.

As well as providing opportunities for grounding, the teaching of principles in examples, practicals and case studies show how a permaculture design framework fits into a larger body of knowledge of practical techniques. I aim to show students what types of information on techniques are available and ways in which they can be accessed. I make it clear in the pre-course information pack that this course is not a how-to course on techniques, eg organic gardening or natural building, but focuses on the process of design, which can be used on a myriad of scales.

Patterns in knowing and design

This process of teaching emphasises connections and encourages systemic analysis. It extends simple ecological principles, eg edge effect, succession, cycles, into new ways of thinking and generating ideas about how humans can interact with the environment. This course teaches methods of design grounded in the principles of strategic sustainability.

Bibliography

1 The Natural Step UK, Thornbury House, 18 High Street, Cheltenham, Glos, GL50 1DZ. Tel: (01242) 262744
2 The Mind Map Book
3 Serious Creativity
4 I am Right, You are Wrong, de Bono, E, Viking, 1990

Further research

The Ecology of Commerce

More information on the work of Joanne Tippett and the Holocene Design Company can be found on the internet at http://www.holocene.net and http://www.geocities.com/RainForest/Vines/8674/dis.html (the latter for thesis).

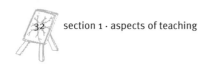

Patterns for Course Design –
An Open Agenda Approach

Andy Langford and Jane Hera

An open agenda approach seeks to put course students in charge by giving them frameworks and methods which they can use to:

a) brief us as course designers, and.

b) as their capacity develops, come to design more and more of the course themselves.

The two main rationales for using this approach are:

· That the course becomes a living design project and thus provides a unique opportunity for us to model working with design frameworks (BREDIM, SADI, pattern languages – see pages 100-102 for explanation of these) and design methods (zoning, sectoring etc). What is special about this design opportunity is that it gets implemented and reviewed each time the class meets and so, unlike land-based projects which take time to mature, people get to see the results of the design in action, reflect on it and get to feedback their observation to us. We tweak the design visibly and attempt to articulate what we are up to. This is currently the least developed aspect of the open agenda approach and we'll be working on least effort for most effect methods for making our thinking transparent. (See 'Articulating the design', below.)

· That this approach goes some way to resolving the dilemma of who is in charge, the teachers or the students. What happens with this approach is that the teachers start out, clearly in charge, and devolve power over time. They move to become facilitators and enablers for the group rather than directors. The potential transfer of power is flagged up at the outset. We find that this means students are pleased to mandate us to be in charge in the early stages, whilst they build capacity for self direction using the structures and methods we provide, and any others they may have at hand.

The gradual shift of power avoids overwhelming people in the early days. Recently we have found it useful to put the design focus on something more contained than the whole course. Two groups we are currently working with are once-a-week classes. For them we are using room layout and functional decoration of the room as an initial practical design project.

Layout and decorations have to be installed and removed in 10 minutes at the beginning and end of class, and the decorations stored in a filing cabinet drawer for use the next week. Functionality is included by having the decorations map the course content as we go. This is a light and fun project that gives people a feel for working up designs in large groups or complex situations.

Using this approach means that we have given up attempts to programme a course in any detail in advance. We like to start with a known programme for the first day or so, and then use the group to generate the next steps. This sometimes offends people who want to drop in on a course to get specific tasty bits. We find it difficult to accommodate such requests, and advise such people that they will have to take pot luck.

However there are usually some significant events, trips to projects etc, which have to be programmed well in advance. The open agenda approach can cope readily enough with these fixed points.

○ **Articulating the design**

Articulating the design thinking that shapes the course is demanding. It requires that we make clear the basis for some of the typically seat-of-the-pants decision making that is involved. It helps if we have described designing for emergent situations early on and have talked about the twin design approaches of rational design and intuitive, or seat-of-the-pants, design. Rational and intuitive design approaches are flagged up when we draw up a recently developed mind-map (see 'Mind-map' below) which attempts to show the whole of permaculture using less than 40 words.

Mind-mapping the agenda for the day or session on a separate white board which remains in view is another way of showing that the event has a design. It also allows us to gently introduce the idea of looking at the design as distinct from the content.

○ **Mind-map**

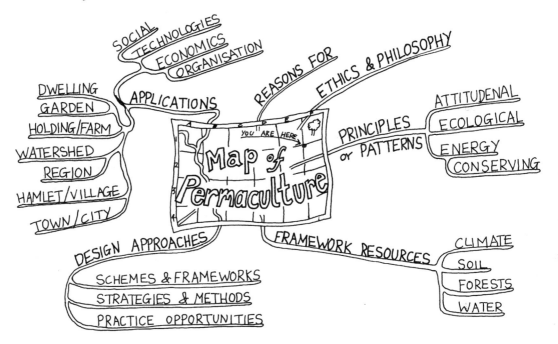

This mind-map is multi-functional beyond describing design approaches:

· It gives us an opportunity to cover, in brief, the wide scope of permaculture. People find this inspiring – it sets the scene for a positive course.

· It provides students with a set of headings which they can use to order their own notes. This is important as the open agenda approach means that we could be going anywhere at any time. We are not following a sequential route as we might have been if using a more conventional, timetabled strategy. Students can rework session material to fit it in with one or more headings given by the menu. This reworking also serves as an opportunity for revision.

· It provides a menu or map for students to use to direct their thinking and, indeed, to direct the course. Whenever there is any hiatus, or if people want a change, we can ask them to choose an item of the menu for the class to address.

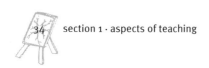

Matters Arising – the pattern that drives the course

We run a Matters Arising session every four sessions or so on all our Design Courses. This is the mechanism that keeps students briefings to us up-to-date.

A Matters Arising starts with people in pairs, where one person thinks aloud for five minutes and the other actively listens. After five minutes they swap. This is called a Think and Listen (Ref 1). The set task for this is to generate questions and observations regarding aspects of permaculture recently covered in the course, or any other aspect of permaculture that is current for the questioner.

Each person gets to offer up one item which is patterned into a visible mind-map. With big groups people can be encouraged to pass. Then we work through the pattern to see what deserves a session of its own and what can be answered there and then. Every topic gets at least a paragraph length mention even if it is going to get full length treatment later. We are always delighted to have material visited more than once as this serves as a review.

We sometimes have someone in the group speak to a topic, or use it as an opportunity to engage the group in a spoken revision or speak to it ourselves.

We take every chance to use this mechanism to open up the range of topics that the group has covered. This way topics that might have taken a whole prepared session can be telescoped into a 10 minute answer in a Matters Arising.

There is no knowing beforehand where such a session will go and how long it will last. Sometimes the energy and attention is good enough to spend two or three hours, with breaks, working this way. It is always great fun and the topics typically range from the micro, eg how do I sow seeds in a mulch?, to the macro, why does agribusiness continue to insist on ploughing?

Feedback Go-Rounds

We frequently ask people to say what is going well for them and what they might have done differently, or what has been difficult for them, perhaps at the end of each day or so. This gives us great feedback about the quality of our design work.

People feedback in a Go-Round. Each person speaks for half a minute or so. With big groups we time this accurately so as to avoid air-hogs pushing the overall timing out. Every third or fourth feedback Go-Round is proceeded by a Think and Listen of three minutes each way, so that we know people have had an opportunity to think before they speak. This way we are sure to get wholesome feedback of great utility.

Using Matters Arising in conjunction with Feedback Go-Rounds usually sets us up with big clues as to what to modify or rethink entirely, and we respond by making changes the next session or day. The rapid response is much appreciated by the students and is clear evidence that we are thinking as designers.

Bibliography

For descriptions of how to run Think and Listens, Go-Rounds, Support Groups and more, see Designing Productive Meetings and Events, Permaculture Academy Occasional Papers No 1, 1998, published by South Oxfordshire District Council, LA 21 Dept, Crowmarsh Gifford, Wallingford, OX10 8HQ. Tel: (01491) 823497.

Do I Teach?

Skye

To be a teacher, is to impose rank. When I stand in front of a class, I am conscious that the students have agreed, temporarily at least, to assign me a higher rank than themselves. I have the knowledge. They do not.

And I am uncomfortable with that. The question of rank is one of the most insidious ills in our society, and yet it is often invisible to us, except in a few obvious ways. That you are reading this, indicates a privileged rank. You could afford to buy the book, or are a member of a library. You can read English. These two facts alone rank you above 90% of the world's population.

As a member of the social elite of the world, how are you going to use your position of power?

You can stand in front of the class and tell them what they must learn. You can give them the information you think they need to know. And they will listen, in the hope that they too can join the elite group, called Permaculture.

Or you sit as a member of a circle, and join in the exciting experience of learning.

I do not teach. I simply join with other people to experience a joint journey into new ideas, new information, new insights. Insights into the world, into Nature, into people and into myself. To be permitted to sit under a tree with a group of campesinos and share a small part of our lives is a humbling experience. I feel honoured to be permitted such an opportunity.

I do not teach, I listen. And if the time is right, I may be given my turn to share some of what is important to me: my hopes; my dreams; my love for Nature; my love and respect and compassion for people; my fascination for interesting technologies that are less damaging to the environment; my desire to eat healthy, tasty food; my hope to live with other people in harmony and peace; my need to be loved and appreciated, and my anger at the injustice in the world.

I have nothing to teach others. I can only share who I am. That is all I can offer. And I hope that others may wish to listen to my story, just as I wish to listen to theirs.

Don't teach. Listen to your own story. Tell it. And listen very carefully to the stories of others. My desire is to always be the student, never the teacher.

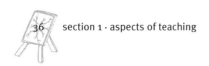

What? Car Maintenance in a Design Course!

Peter Harper

Editor's note:

During initial consultations for this book, Peter put forward Car Maintenance as a topic he would include in the Design Course. Peter has been writing and teaching about sustainability for many years, and so I asked him to write a piece that would explain why he would include Car Maintenance. It is therefore different to other contributions in this section, and is intended to be deliberately and creatively provocative.

Introduction

What is the Design Course for? Seventy two hours is not very long, and we cannot expect to get very deeply into everything. I can think of three reasonable responses to the shortage of time.

- Sampler. Its aim would not be to actually teach anything, although that might happen, but to present information and set students off on a lifetime of exploration.
- Unique features of permaculture. Since we only have 72 hours it would make sense only to teach things that students cannot find elsewhere, material unique to the Design Course.
- Introduction to modern sustainable living, accurately reflecting the real world situation.

These are not mutually exclusive, but I would like to argue for the third alternative on the grounds that it is desperately needed and the Design Course is the closest thing we have to it.

In my view the principal aim of permaculture is to help humanity through the 21st century by developing models of efficient, materially lean but culturally rich modern living. A great part of this will come about by training a proportion of the people in the richer parts of the world to reduce their ecological footprint down to 20% (Ref 1), perhaps less, of the prevailing levels, not just by technology but by much deeper and subtler changes in culture, lifestyle, tastes and organisation.

If we start with permaculture ethics, Limits to Consumption informs the whole course: it is an absolute cornerstone and distinguishes us sharply from the mainstream approach to sustainability. The 72 hours can then be equally divided between Earth Care and People Care, although of course there are lots of overlaps and fuzzy areas. I think the Course as commonly taught does People Care quite well and I shall say no more about it.

Ecological Footprint Analysis

This is an accounting tool, which converts resource consumption and waste assimilation into land area. All activities require the use of various types of finite natural resources, and these can be converted into land-area equivalents. The area of land needed for the combined categories of consumption and waste disposal generated by a given community is known as its ecological footprint. If a community's ecological footprint is larger than its physical land area, then the difference is known as acquired carrying capacity – or ghost acres.

Converting consumption into land area – the production and use of any good service depends on various types of ecological productivity. These ecological productivities can be converted to land-area equivalents. Summing the land requirements for all significant categories of consumption and waste estimates the EF for the reference population.

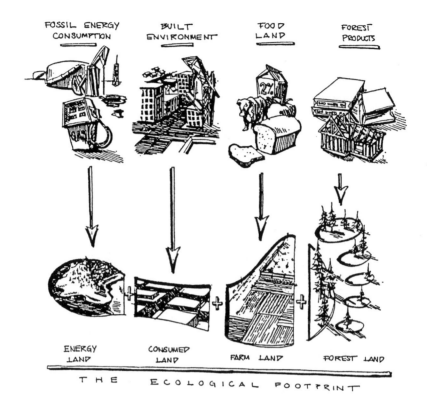

Determining the footprint of commuting

References and assumptions

The US has 15 million hectares of roads, mainly for cars (and mostly built on agricultural land).

There are 1.75 million people per car in the US.

We assume 230 workdays per year.

According to the Bicycle Federation of America, a bicycle rider requires 900 kJ of food per 10 kilometres.

According to Environment Canada, cars make up to 98.4 per cent of the traffic in Vancouver rush hour, while carrying only 62 per cent of the commuters. Hence, we can conclude that a bus passenger needs only about 2.6 per cent of the road space that a car driver occupies. (For mathematical wizards, the calculation is as follows: (0.016/38)/(0.984/0.62)=0.026)

Calculations

Bicycle: The bicycle rider requires an extra 900 kJ/day of food for this daily 10 kilometre trip. We assume that this extra energy stems from breakfast cereals. These cereals need land to grow and energy for processing. The land equivalent of the commercial energy needed for agricultural production and for food processing of plant crops is typically the same as the crop area; hence, the total land area for the growing and processing of the food is double the growing area. The road space is assumed to be neglible. Cereals have a nutritional content of about 13,000 kJ per kilogram. The world average in agricultural production is 2,600 kilograms of cereals per hectare per year.

$$\frac{900\,[kJ/cap/day] \times 230\,[days/yr] \times 2}{13,000\,[kJ/kg] \times 2,600\,[kg/ha/yr]} = 0.0122 \text{ hectares or } \mathbf{122\ square\ metres} \text{ per rider}$$

Car: The average direct gas consumption by American cars is about 12 litres per 100 kilometres; indirect carbon consumption for car manufacturing and road maintenance adds 45 percent. Each litre of gasoline contains about 35 megajoules or 0.035 gigajoules of energy. Therefore, the fossil fuel Footprint of auto commuting is:

$$\frac{1.45 \times 12\,[1/100km] \times 0.035\,[GJ/l] \times 10\,[km/day] \times 230\,[workdays/yr]}{100\,[km] \times 100\,[GJ/ha/yr]} = 0.14\,[ha/cap] = 1,400\,[m^2/cap]$$

In addition, cars need road space. The road space per US citizen is:

$$\frac{15,000\,[\text{ha}]}{259,000,000\,[\text{Americans}]} = 0.06\,[\text{ha/cap}] = 600\,[\text{m}^2/\text{capita}]$$

Cars use 97.4 percent of the road space. However, the daily (2 x 5 [km] =) 10 [km] commute represents only about 1/8 of average annual car usage, and every car represents 1.75 people. Therefore, the per capita road space required for the 10 kilometre commute would be (0.974 x 1/8 x 1.75 x 600) = 128 [m²], mostly on agricultural land. Therefore, the total land appropriation for single occupancy car commuting sums to **1,530 square metres** of land.

Bus: The energy requirement of short-distance buses is 0.9 [MJ/cap/km]. Indirect energy requirements for roads, buses and maintenance are assumed to add an additional 45 percent (same as cars).

$$\frac{1.45 \times 0.0009\,[\text{GJ/cap/day}] \times 230\,[\text{workdays/yr}] \times 10\,[\text{km/d}]}{100\,[\text{GJ/ha/yr}]} = 0.03\,[\text{ha/cap}] = 300\,[\text{m}^2/\text{cap}]$$

In addition, buses need road space. As a first approximation (see above) we assume that a bus passenger would only use 2.6 percent of the road space a car driver requires for the same distance, ie (128 x 0.026 =)3 [m²]. Therefore, the total land appropriation for the two daily 5 [km] bus rides requires approximately **303 square metres.**

Source: Our Ecological Footprint

The background assumptions

If the Course is run as an introduction to Sustainable Living, there remains the problem of defining sustainability. It's a fine-sounding word, easy to bandy about (like permaculture?), but what you do about it depends on all sorts of assumptions about how the world works and what is going to happen in the next hundred years or so. Views on sustainability can differ quite a lot. Mine are unusual in permaculture circles in that I believe sustainability requires an urban, high-tech future, not a rural, low-tech one, albeit with limits to consumption. There is no going back to peasant culture. Modern Life is here to stay, and will be worldwide by the end of the 21st century.

I believe the permaculture movement can have an important role, but we must play it right. We are not the whole solution, only part of it, just as vitamins are essential but not a whole diet. We must not try and insist that everyone lives on pure vitamins. Paradoxically perhaps, most of the work on sustainability will be done by the mainstream processes of governments, markets, industry, diplomacy, mass society and big technology. This works because they go with the grain of 'development' and consumer aspirations.

Let me give an example. Dry or composting toilets make a lot of environmental sense and rank high in the pantheon of green technologies. They use no water, and may save 30% of your household water consumption in one jump, not to mention various other benefits. Hooray! On the other hand, take-up is likely to be patchy even with the strongest encouragement. I'd be surprised if more than a few hundred thousand British households were both willing and physically able to install a dry toilet. Say a million, that's about 2% of the population. The overall effect is then 30% x 2% = a 0.6% reduction in national water use. Contrast ultra-low-flush toilets which deliver a reduction of 'merely' 20% of household water consumption. You may argue that these are a pathetic bourgeois cop-out, but follow the numbers through: since these toilets are almost indistinguishable from the standard pattern it is quite plausible to suppose that they could become standard and gradually replace 50% or more of existing toilets. The net effect would be 20% x 50% = 10%, sixteen times better. Which should the Earth Care ethic prefer? Perfectionist, gung-ho-greenie solutions are not always the most effective in reducing total environmental impact.

I can multiply such examples endlessly. We must accept that we can often be outperformed environmentally by the mainstream approach, so let the mainstream get on with what it's good at. We should remain faithful to our distinctive task of articulating the Limits to Consumption ethic. In the

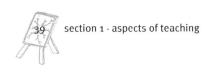

longer term the whole question of Limits to Consumption will be essential to sanity and even survival. A difference between us and the mainstream is that we are culturally much more flexible, and free to explore a far greater range of options, some of which may turn out to be a vital part of the solution in ways that we cannot yet guess.

Additionally, there should be no such thing as 'urban permaculture' because it should be fundamentally urban anyway. That's where everybody lives, and in general dense settlements are much more efficient at providing the needs of human beings. Living in remote locations is usually a step backwards in sustainability terms.

Three principles

I would like to mention three principles which householders wishing to become sustainable would do well to consider, and which could be taught as part of the course. They are similar in some respects to the design principles but more fundamental.

1 Numbers

The toilets example I've just given shows how a few ballpark figures and some elementary calculations help us to make judgments and decisions. Numbers aren't everything of course, and should be overridden by other considerations if necessary. But we must have at least approximate numbers to overcome illusions of appearance or wishful thinking; to demonstrate why one course of action should be preferred to another; to save ourselves a lot of trouble doing something ineffective or even going in the wrong direction; and as a gauge of our progress towards greater sustainability. Therefore we need to be able to deploy basic statistics and numerical justification for pursuing path A or path B in preference to others. These are useful skills to pass on to students.

2 Don't aim for perfection

Generally speaking, in any one area of sustainability – energy, water, waste, transport etc – you can reduce your footprint by 10 or 20% with very little effort or cost, just sensible habits and management measures. Fifty per cent usually takes a fair bit of investment but is worth while. On the whole it is better to call a halt here and attend to the quick-and-easy measures in some other area, and so on. Then go round again. The crucial thing is not to get obsessed with any one area, because it gets harder and harder as you get towards perfection and gobbles up your time and resources.

Example: having showers instead of baths could save 15% of your water for nothing; putting a 'hippo' (or brick, or squash-bottle) in the cistern saves another 5% of household water for a negligible outlay. Low-flow taps and shower heads could get you to 25% for an extra £20 – less than £1 for each % saved. A super-low-flush toilet instead of a hippo would cost £180 with a total of 40%: £5 per %, getting less cost-effective but still reasonable, and a larger total. A commercial grey-water recycling system might save 25% but will cost around £1500 to install, £100 a year in running costs and endless technical headaches: over £60 per % saved. Therefore, in spite of the attractively perma-kosher ring of greywater recycling, it is not the thing to do at the beginning. The money is far better used in other directions first.

I call this the Quit While You're Ahead principle and it often crops up in various guises in Design Courses. Although as a broad aim we are looking to reduce our footprint by 80%, 50% is quite good enough to be getting on with in each area unless progress proves to be ridiculously easy.

The ability to deploy sequences of measures ranked in order of cost effectiveness is another valuable skill for sustainable householders and Design Course teachers.

3 Sorting the wheat from the chaff

In order to make a serious contribution to the sustainability challenge we have to check things out carefully and critically and be prepared to junk what turns out to be wrong. At this stage of the game

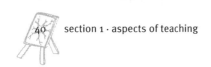

I would argue that it is more important for us to generate results and communicate them to each other than actually to be personally ever-so green. There are so few of us that our personal eco-rectitude makes a negligible difference physically, but clear results widely reported could amplify our efforts thousands of times. A result then, is worth a thousand cabbages.

I would like a session in the Course to deal with setting up controlled experiments, writing up the results and reporting them. This way we can gradually build up the body of reliable knowledge. Doing careful experiments can actually take quite a lot of time, so it is impossible to cover more than a few areas. Students on the Course could be encouraged therefore to pick one bite-sized area of specialisation for a first go at original research. It can be very exciting.

Content of the Earth Care bit

In teaching sustainability issues myself I have found it helpful to break household activity into various categories. Sometimes I do it in functional categories like this:

Heating

Transport

Household repairs and maintenance

Food, including buying, preparation and clearing up; might include some garden produce

Leisure and holidays, includes gardening

Clothes and laundry

Electricity and appliances

Personal washing and hygiene, a lot of water stuff

Work

Shopping.

Sometimes I break it down by the flows of energy and material through the household system:

Energy

Water

Food and nutrients

Solid waste

Materials.

Whichever way I do it I divide the material in each category into several sections (with approximate time proportion):

· Global background, typical patterns of generation and consumption, major problems (20%).

· Principles: basic jargon and statistics, how they work, any controversies or uncertainties, the mainstream approach (20%).

· Footprint reduction, how-to, examples, skills needed, practicals, ideas for experiments (50%).

· Tools, references, examples to go and see (10%).

The raft of skills needed for an 80% reduction of footprint on your own is quite substantial. We're talking degree-course stuff here, many years of learning and experience. The 72-hour course is merely an introduction. We need to decide how much time can be devoted to actually teaching certain skills, and how much to simply identifying ones that need to be acquired, and pointing people in the right direction.

As to the relative time to be spent on actual topics it's probably sensible to relate this to their significance in environmental impact, and how much the householder can do to influence them. For typical households the ranking is probably:

Transport ╌╌⟩ Household Energy ╌╌⟩ The Food System ╌╌⟩ Purchasing Choices ╌╌⟩ Water ╌╌⟩ Solid Waste

Each of these breaks down into lots of categories which a Course could look at, and of course they overlap a lot. There are two other things which correlate strongly with the footprint of a household: number of children and income. They are culturally sensitive but must be addressed. It is as difficult for a very rich person to be sustainable as for a camel to pass through the eye of a needle – unless the surplus is carefully invested in increasing sustainability.

· There has to be some overall perspective on the global situation. There usually is. I would also like to see an attempt to map the 'landscape' of sustainability activity – who's doing what and where we all fit in. Which parts of the landscape are actually permaculture under another name? What are permaculture's immediate neighbours? Where would you go for specialist information on x or y? Who might we fruitfully collaborate with?

· The footprint implications of energy are usually ranked 10-20 times greater than things like water or solid waste, so it deserves much more attention than it is usually given.

· Transport. Car energy is as much as household energy for many households. If you don't have a car you're already miles ahead of nearly everybody else. If you do have one you will never get to 20%, and you will be embroiled in the oddity that small improvements on the car can outrank many more eco-sounding actions elsewhere. For example, say your car uses 30 gigajoules a year. Good tuning, tyres, cap and leads, timing, plugs, taking off the roof-rack when not needed, good road atlas so you don't get lost (say £50 extra a year) could improve performance by 10% – saving 30GJ over 10 years, at a cost of £17 per GJ saved. Compare a commercial solar water-heating system, which might save the same amount at an installation cost of £3000 – £100 per GJ saved which is over five times less cost-effective. Weirdly, showing people on a Course how to do some basic car maintenance – clean plugs, change and set points, change oil and air filters, check tyres and battery – could lead to greater environmental improvements than most other things they could do.

   The environmental cost of an annual holiday in the sun is also a big one. A return flight to Majorca uses 10GJ, which is similar to the average annual energy cost of water heating. A single to Melbourne is 72GJ, equal to all other annual household energy. A quarter of the liquid fuels used in London is aviation fuel at Heathrow! All this for perspective: it makes little sense to bust a gut trying to save a few megajoules here and there, for example not using a motor mower, when you blow ten times as much in one throw and hardly notice. Look at the whole system and go for the jugular.

· The food system. This has classically been central to permaculture and the Design Course, and when people speak about self-sufficiency they mean principally food. The food system is a major contributor to footprint; a recent Danish study of household activities rated it above both transport and house heating. Yet for the typical British household food accounts for less than 15% of total expenditure.

A lot of the Course has traditionally been devoted to food growing and gardens. If this is supposed to reflect its importance in footprint reduction, it is a mistake. I estimate that changes of diet and food purchasing are 10 times more cost effective than growing your own, which is a lot of fun but a tremendous time-waster.

Let me hasten to say I am professionally a gardener. I design gardens, create them, write about them, and I grow a lot of veggies. Gardens are great, but they take up too much space in a Design Course if we are really trying to reduce footprints and teach others how to do the same. Gardening is inspiring but as traditionally taught does not reduce your footprint much because your footprint is coming from somewhere else. So is your diet. The value, net of actual costs, of the fruit and veg from a typical

garden is unlikely to be more than 10% of what you spend on food, or more than 2% of your household income. Of course there are other 'yields' but economically they are small potatoes: they are chiefly cultural. If you have an allotment or larger piece of ground you may be able to increase these figures somewhat, but the amount of time it takes can seriously erode your other footprint-reducing activities. There should be a big 'health warning' about growing your own.

Assuming you get 90% of your food from elsewhere, diet and purchasing policy will have a far greater effect on your footprint, and the take-up of new dietary and food-choice ideas is likely to be far, far higher than growing more of your own. Basically you:

· Eat less meat, or none at all.
· Eat less processed stuff.
· Eat local wherever possible.
· Eat organic wherever possible.
· Eat ethical wherever possible.
· Eat as large a proportion of your diet as possible in the form of fresh fruit and veg.
· Most of the rest as complex carbohydrates.
· Support local growers or box systems; offer to help them.

In summary

I would like to see less time in the Design Course for gardens and land use. Of that which is allocated, I'd like to see less spent on growing your own food and more on other uses of gardens, such as waste reclamation and recreation.

Conclusion

You may have found my opinions and arguments rather surprising. People will ask you to refute them. Let me help you here.

· Remember that my proposals only apply if we are trying to make the Design Course accurately reflect the priorities of modern, urban, sustainable living.
· They assume we are trying to make mainstream society sustainable rather than create a separate sustainable culture.
· This might not be the best role for the Design Course (see my introduction).
· I interpret 'sustainable' in an unusually harsh and quantitative fashion, and this can be dismissed as eccentric.

However, it is probably fair to say that the kind of approach I have taken is gaining in strength in the wider environmental movement, and will exert an influence on permaculture theory sooner or later. We might as well be ready for it!

Bibliography

1 Our Ecological Footprint

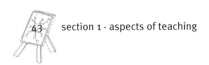

How Do I Become a Permaculture Teacher?

Patrick Whitefield

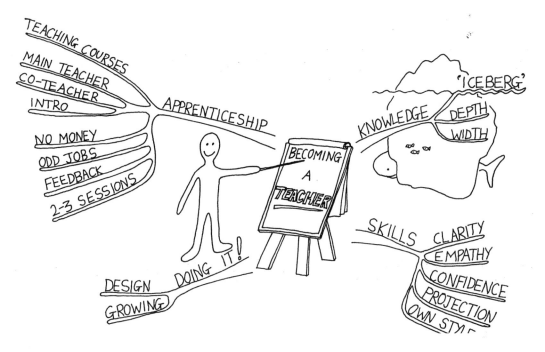

There is no recognised pathway to becoming a teacher of permaculture Design Courses, nor a specific qualification which entitles you to teach them. You must have completed a Design Course yourself (see 'Introduction'), but beyond that it's left to the integrity of the individual teacher.

This kind of self-assessment is no light task. Permaculture itself is an excellent concept, and both students and the Earth herself deserve that it is taught excellently. There are two aspects to excellence in teaching: knowledge and teaching skills.

Knowledge

The knowledge required for good teaching is like an iceberg: nine-tenths of it is invisible at any one time. In order to teach well it's necessary to have a depth and breadth of knowledge far in excess of what you are able to put across in the time available on any one design course. There are three reasons for this.

Firstly, it gives a greater depth and resonance to what you're saying; you understand the 'why' of what you're saying as well as the 'what'; you can choose from a wide range of material to fit the needs of individual teaching situations; you can give a wide range of examples, to bring the principles to life and make them both practical and relevant to the students.

Secondly, it means you can answer questions on a wide range of subjects, and if there is no time to answer in full, give students a clear guide to where they can find out more; you can cover subjects you hadn't intended to if the group asks for them. You can't expect to know everything, but there's a big difference between an occasional 'I don't know' on the one hand and knowing little more than the core syllabus on the other.

Thirdly, you can approach the standard body of permaculture ideas with an informed skepticism. Much harm has been done by teachers repeating stock ideas from the permaculture books without any assessment of how suitable or relevant they are for the people being taught. In particular, things

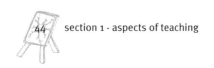

which may be widely applicable in Australia may be of limited relevance here. A common example is an unthinking acceptance that swales are always a good idea.

In practice this means that the knowledge gained from a single Design Course, several Design Courses, or even several Design Courses plus reading all the permaculture books, is unlikely to be enough.

One solution is to draw on one's existing skills. Many people come to permaculture with skills both theoretical and practical in a related field: farming, gardening, forestry, building, or landscape design for example. Personally I have found my theoretical training in agriculture and practical experience in farming, gardening and nature conservation essential parts of the total package I bring to my teaching.

Alternatively these skills can be acquired after one's introduction to permaculture. I would certainly recommend to any prospective teacher that they get some detailed training or experience in one of the earth-based skills before taking on the role of main teacher on a Design Course.

Background knowledge in depth is only part of the story. A Design Course teacher also needs to have a wide knowledge of all the subjects which go to make up permaculture. Since I took my Design Course and started looking for relevant information I've found the world is full of it. I'd never noticed it before because I wasn't looking for it. Now I spend as much of my time as possible reading journals and books on related subjects. In particular I make sure to keep up-to-date with new ideas and information. It's all too easy to become outdated.

Of course it's not necessary, or even possible, for every teacher to have detailed knowledge of every aspect of permaculture. It's far too wide-ranging for that. In fact different teachers are known for their specialities, and students may be attracted to a specific course because the main teacher is a specialist in, say, woodland. But it's highly desirable that every major subject on the course should be taught by someone with a good background in it. This may mean bringing in guest teachers. For example, on my courses I always try to have some extra input from someone who knows more about building than I do.

In general, teaching near the limit of your knowledge is unsatisfactory. I find I can get away with it for the occasional session on a minor subject – in which case I always tell the group what I'm doing. But a whole course taught like that can leave students with something insubstantial and unrelated to real life; an experience that is of little use to them, practical or otherwise.

Teaching skills

It's often assumed that if someone knows about a subject they can teach it. But teaching itself, like any other activity, is a skill.

Some people are born with a natural ability to teach, which they can enhance by training; others can learn how to teach; others have talents which lie in different fields, and would be well advised not to try teaching. Every prospective teacher has a responsibility to carefully consider where they lie on this continuum and act accordingly. A good way to find out is to do a few trial teaching sessions with feedback from as many different people as possible.

There are four main skills involved:
· Clarity
· Empathy
· Confidence
· Projection.

Clarity is the ability to transform the seamless experience of reality into a structure which is clear, definite and easy to put across. (Of course you need to remind students that what you present is not

actual reality. That's infinitely more complex.) It's also about prioritising, deciding which things are the most important to include in the limited time available.

Empathy is, amongst other things, being able to put yourself in the position of the person who doesn't already know what you know. It sounds obvious, but it's surprising how many people attempt to teach without really being aware of this. Very often these are people who are the best at actually doing it – who famously tend to make the worst teachers.

Most teachers experience some nervousness in one form or another. But it varies in degree, and there's a difference between feeling a lack of self-confidence and showing it. A teacher who is obviously unconfident in their presentation will not inspire confidence in what they are teaching. Self-confidence can grow with experience.

Projection includes the ability to speak clearly, a sensitivity to what's going on in the room, and the ability to react to that in ways which will enhance learning. It also includes an element of theatre. Teaching is an art as much as a science. Perhaps most importantly of all it includes genuineness, the quality which is evident when we speak direct from the heart about something we care about deeply.

Every teacher needs to develop their own style. There are many different ways of teaching permaculture and it would be a rash person who said that one is more valid than another. Each person will find that one particular style suits them. It may be modelled on that of another teacher or it may be what you have developed yourself.

Whichever it is, the important thing is that your style fits your talents and personality as closely as possible, that it genuinely reflects who you are. There's no point in adopting a theory or another teacher's style, however admirable, if it doesn't fit you.

Apprenticeship

The best way to learn teaching skills is on the job. Local adult education services run courses for teachers and these are certainly worthwhile. From time to time permaculture teaching courses are run, and these can be valuable too, as long as you remember that they always represent one way to do it, not the way. But the best way to start is by working as an apprentice to an established teacher.

Apprenticeships vary, and there is no norm for the agreement between the two parties. But I would suggest the following points are helpful.

· The apprentice teaches two or three sessions on the course. They should be on subjects you have a depth of knowledge about – remember the iceberg effect. Videoing one or more of these sessions can be very valuable to the apprentice.
· Teacher and apprentice have a regular feedback session each day where each feeds back to the other about their work. The apprentice is encouraged to observe every aspect of the course – teaching, timetabling, administration, domestic etc – to consider how well things are working, and why.
· The apprentice is available to do odd jobs, such as photocopying, collecting people from the station, helping with the cooking if necessary.
· No money changes hands, except that on a residential course the apprentice pays for his/her food and lodging.

It's beneficial to apprentice on a number of different courses, preferably with a number of different teachers. The next step is to co-teach with one or more other teachers, perhaps on a weekend introductory course before embarking on your first Design Course.

Some teachers always co-teach, others go on to teach alone when they feel their skills and experience are up to it. In fact no good teacher teaches entirely alone. A number of guest teachers are necessary, both to give variety of presentation and to cover areas where the main teacher is less knowledgeable. Neither approach is better than the other; it's a matter of what you feel comfortable with. But if you do take on the role of main teacher, your personal responsibility to provide the level of excellence which permaculture deserves is that much greater.

The process of apprenticeship never really ends. There's always something new to learn, however experienced we may become.

Doing it

Teaching permaculture without actually doing it is a pretty hollow activity, and this hollowness comes across in the quality of the teaching. Doing it includes both design work and actual physical work on the ground – growing things.

Most of us probably did a rash of designs after completing our own Design Courses, perhaps in working towards the Diploma. But it's equally important to go on designing while teaching. If a student during a design exercise asks 'What would you do in this situation?' an answer based on actual experience is worth a thousand drawn from theory.

One problem can be the lack of demand for design work. Most people who are interested in permaculture are not the type who want a consultant to come round and tell them what to do. They'd rather go on a course, learn about it for themselves and do their own design. One way round this is to offer to design people's places for them for no charge. Many people are keen to accept such offers. But this is rather a second best to designing for someone who positively wants it. Whether money changes hands or not, the latter is a more realistic situation.

Some of us have more opportunity to work on the land than others, whether for reasons of family commitments, access to land, health or whatever. But it must be a rare person who has no opportunity at all, even if it's just helping a neighbour occasionally, or making the most of a tower-block balcony.

Growing things is not only valuable in itself, but is an essential part of the teacher's job. The authenticity which flows from knowing permaculture not just with your heart and mind but also with your body is a unique and irreplaceable quality.

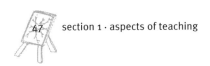

Permaculture Association (Britain) Teaching Policy

Association Council of Management

Becoming a permaculture teacher

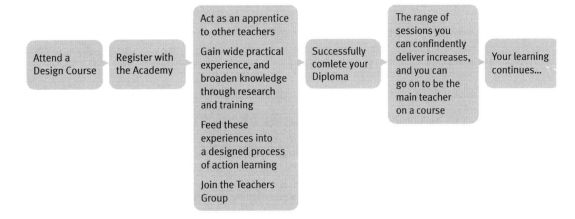

Attend a Design Course → Register with the Academy → Act as an apprentice to other teachers / Gain wide practical experience, and broaden knowledge through research and training / Feed these experiences into a designed process of action learning / Join the Teachers Group → Successfully comlete your Diploma → The range of sessions you can confindently deliver increases, and you can go on to be the main teacher on a course → Your learning continues...

Who can teach?

· Only those who have graduated from the 72-hour Permaculture Design Course are eligible to teach permaculture.

· Teachers should have two years of practical experience before taking on the role of lead teacher within a course. Gaining a Diploma is clear evidence of this. However graduates can take on an apprentice role during this time. It is therefore recommended that those wishing to teach find an experienced teacher who is willing to provide apprenticeship or training.

The teacher's responsibilities

· To ensure that course participants are members of the Association. This can be done by incorporating the cost of membership into the course, with discounts for students who are already members. This is strongly encouraged, as this links new graduates into a network of people who are learning about and doing permaculture, and gives them greater access to research and contacts. Membership can thus be considered as a vital element of after-course support.

· To send details of graduates to the office, so that they can be entered into the graduates database.

· To inform the office of courses run, so that accurate records of numbers of courses and geographical spread can be established.

· To ensure that their courses are open to all, with no discrimination on the grounds of age, sex, culture, religion or education.

· It is recommended that courses should provide at least one free or reduced fee place for those on low incomes.

The Permaculture Association will support teachers by:

· Co-ordinating a Teachers Network with regular training events and opportunities for teachers to gather and develop solutions to common problems and share best practice, materials and skills.

- Providing details of courses to Association members via the newsletter, and on request.

- Providing course certificates.

- Providing after-course support for students, such as useful contacts, details of projects in their area and answering enquiries.

- Holding details of members wishing to do courses, and passing these to local teachers.

- Holding up-to-date records of teachers and graduates, for use by projects and courses.

- Working to raise the profile and desirability of permaculture courses to the general public and specialist groups.

○ Additional requests from the Association

- If you are teaching Design Courses, please let us know. We will add your name to the Teachers Group mailing list, and will be able to put prospective local students in touch with you.

- There is a great need for further outreach. This applies to both geographical spread and the range of participants. We hope that the good work done by Patsy Garrard and George Sobol in developing links with the Workers' Educational Association (WEA), and others working with Further and Higher Education, NVQs and SNVQs, will go some way towards making the course more accessible. The Association would welcome proposals or plans that would further help to deliver permaculture courses to groups and individuals that are usually excluded, for whatever reason, from participating in education or training.

- There is also a need for more teachers from a wider range of backgrounds and cultures. This can be greatly helped if teachers can try to spot and encourage enthusiasm within students and graduates. If you are willing to provide apprenticeship to prospective teachers, please let the Association know.

- Sharing materials. This guide has brought together a wide range of teaching materials, but by no means all of them. We hope to continue collating teaching materials and processes. If you have anything you feel would be of interest to other teachers, please send it to the Association. We will make it available to others through the Teachers Group.

- In an attempt to reduce duplication, we ask that teachers who are developing links to Further or Higher Education, keep the Association informed, so that others attempting similar projects can contact you and learn from your experiences.

○ Guidance on awarding certificates

There are two considerations to bear in mind before awarding certificates to students on a course:
- Have they attended at least 70% of the course?
- Did they participate in the main design exercise, and design presentation?

If the answer is yes to both of these questions, then a certificate can be awarded. If the rate of attendance has been lower than 70% of the course, then a certificate can only be awarded if the student can demonstrate that they have put in extra time to catch up. Certificates should not be awarded to those who have not participated in the main design exercise and presentations, as this plays a key role in bringing together the learning built up over the course.

Students who are unable to participate in the group design project and presentations can be encouraged to join another course to complete this. It is helpful if the teacher can help to arrange this.

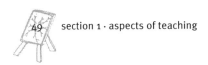

Permaculture Courses and the WEA

George Sobol and Patsy Garrard

Introduction

We have been co-teaching permaculture courses in co-operation with the Workers' Educational Association (WEA) since 1993. Since the first course, run as an evening class, we have also run weekend introductory courses, and the full Design Course as a series of weekends and as the two-week intensive. The WEA has been happy to accommodate any format.

It was through the WEA that in 1997 we submitted the Design Course for accreditation with the Open College Network (OCN) making it a nationally recognised qualification.

What follows describes our experiences. You will need to contact your local Tutor Organiser to confirm details for your own WEA District.

Benefits of working with the WEA

The WEA will:

- Publicise your course in their publicity, through brochures, posters and handouts, selective advertising and direct selective mailing.

- Accept bookings and payments and send out preliminary information for your course.

- Photocopy handouts for your course.

- Offer reduced tuition fees to course participants receiving benefit. Currently course fees in the SW District are £1.70/hour full fee or 55p/hour reduced fee.

- Offer the OCN-accredited Design Course. This course is substantially subsidised for participants on benefit. Contact WEA, SW District, for more details.

- Pay the tutor approximately £16.00 per hour. Pay claims are dealt with promptly on a monthly basis.

- Possibly pay two tutors if you have more than 20 participants.

- Pay travel expenses.

- Provide a venue and audio visual equipment if needed.

- Give the course tutor public liability insurance.

- Enable you as a WEA tutor to attend the City & Guilds Further & Adult Education Teachers Training Course at no cost, if there is one being run by your District. Highly recommended!

- Basically the WEA will deal with the convening of the non-residential side of courses.

What you need to do

- Contact your local WEA Tutor Organiser. If you wish to run the OCN-accredited version of the Design Course ask your local Tutor Organiser to contact the SW District office on (01752) 664989.

- Complete a course information sheet. (Contact the Permaculture Association (Britain) for a copy). This sheet will, among other things, state the Aims and Outcomes for your course. Keep these as

clear and simple as possible. Please send a copy of your aims and objectives to the Association office.

- Secure a minimum number of participants to make the course viable (10 in our district). Payment status of the participants, ie reduced rate or full, should not be relevant.

- Get the course participants to fill in an Enrolment Form at the beginning, and a Learner's Outcomes Form and a Student Satisfaction & Progression Form at the end of the course. You will need to allow time for this form filling.

- Possibly collect some of the tuition fees and forward them to the WEA.

- Keep a register and complete an analysis of the forms at the end of the course.

- If you run the OCN-accredited version of the Design Course you will also need to:
  - complete further paperwork
  - arrange for your moderator's visits to assess your performance, meet the participants, view their journals, etc.

Some other thoughts

- Co-tutoring is a very important model to use on courses. Having more than one teacher brings in the permaculture principle of Each Important Function Is Supported By Many Elements, and preferably creates gender balance.

- Where we have had more than 20 participants on a course the WEA has processed the course as two courses and so paid both tutors. Otherwise we have split a single tutor's fee two ways.

- The format of the course is up to you. We have found that any format is possible, and in theory courses abroad, for British participants, could also be run via your local WEA.

- If you are providing a venue for the course, it may be possible to get venue costs from the WEA. The venue should have public liability insurance.

- If your local Tutor Organiser would like to speak to a colleague who has had experience of permaculture courses, please ask them to contact:

Tutor Organiser for SW District – 01752 664989.

If these notes have failed to answer any important questions, please let me know so that I can cover them in the future. Please feel free to contact us via the Permaculture Association office – 0845 4581805

We hope your association with your local WEA is as fruitful as ours has been.

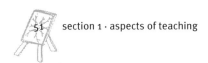

Links with Further Education – My Own Experience

Patrick Whitefield

I have been invited to teach permaculture in colleges both by students and by staff. An invitation from students may come if they want some permaculture input on a related course, such as landscape architecture, which has an 'elective' element in it. But there's no particular reason to suppose that students on the same course in subsequent years will make the same request, so these are usually one-offs.

Work offered by a member of the college staff, on the other hand, can become regular. But you do need to be invited in. I have tried making offers of permaculture teaching to various colleges, addressed to appropriate staff members who have been recommended to me as likely to be sympathetic, but I've never had a response.

The great majority of my work in further education has been for the Worcestershire College of Agriculture, now Pershore and Hindlip College after merging with the Pershore College of Horticulture. I was originally invited to teach at the college in 1991 by Andy Daw, the organic lecturer and a keen permaculturist. Together we taught Forest Garden and Design Courses there. These were funded entirely from the fees paid by students. Gradually I took over most of the teaching on these courses while Andy concentrated on his other work.

One of the fruits of his labour was the Sustainable Land Use Course. He designed this to fit an existing Further Education format, the Continuing Education Certificate, now renamed the Professional Development Award. It's a modular course, with the Permaculture Design Course forming two of the modules. The other subjects include: Soils, Organic Horticulture, Environmental Interactions, Book Study, Project and Sustainable Land Use (a module within the course as well as the name of the whole). To complete the course students must do some independent study in addition to the college tuition.

The college receives government funding for the course, but students still have to pay fees unless they are exempt. Exemption is granted to those who are either: unemployed, on means-tested benefit, eg housing benefit, or under 19. In practice these categories account for most of the students who enrol.

This means that permaculture education can, firstly, lead to an officially recognised qualification, and secondly, be free to at least some people on low incomes.

What's more, students don't need to take the entire course. They can just take the two permaculture modules. In fact the College puts these on as a separate course, ie a normal Design Course, at a different time of year. Due to lack of space at the college campus we hold both Sustainable Land Use and Design Courses at Ragmans Lane Farm, which gives the students the advantage of being taught at one of the best permaculture sites in Britain.

We held the Sustainable Land Use Course for the first time in early 1997. Later that year funding for Further Education was cut by 14% across the board nationally. Some of the courses taught at the college had to go, and Sustainable Land Use and Permaculture could have been among them. Fortunately they were not. I'm not privy to the Principal's reasons for making this decision, but one thing he obviously values about these two courses is that they represent the sum total of what the college is doing about sustainability.

Perhaps this is more of a fig leaf than a genuine attempt to save the planet. But it does provide a unique opportunity for students who do want to spend their lives working towards a more sustainable future. As far as I am aware there is no other course in Britain which covers the same ground as the Sustainable Land Use Course in the same practical way.

Nonetheless it's not oversubscribed, even though recruitment is national. As yet there's not enough demand to make this course viable at more than one college. But in the long run I feel it could become a model of how permaculture teaching can be integrated into mainstream education.

Selling Your Wares

Maddy Harland

Some thoughts on planning

Most people who teach permaculture design view teaching as only one string to their bow. It is rarely the main source of income. For some, it is a valuable experience and a weekend activity – there is still a 'day job' to rely on. For others, it is a part-time, seasonal activity and is mixed with other freelance jobs. If the latter is the case, there will come a time when it will be necessary to register as self-employed. When you do, it is useful to sit down and design a business plan and organise your thoughts. Its purpose is to help you set down everything that is involved in your venture. It will be a blueprint of your ideas and a point of reference to check on day-to-day decisions against your stated objectives. It is also a useful document to draw upon whenever you discuss finances with potential funders, bank managers, business partners, your accountant and anyone else interested in your welfare and plans for the future. The business plan can include your skills, experience and qualifications.

Here are some ideas for questions to ask yourself:
· What type of business is it?
· What other activities will you undertake to earn a living when you are not teaching?
· Do they complement your teaching?
· What is the service?
· Do you intend to start a new business?
· Are you a sole trader or a partnership?
· Who do you see as your customers?
· What will you charge?
· What do your colleagues in equivalent fields charge?
· Are you setting up in an area where there are others offering a similar service?
· Can you co-operate?
· Are you offering something different?
· Consider your place of work: hiring rooms/facilities for teaching.
· Do you have enough money to survive on until your work builds up?
· Should you take every piece of work or should you be more choosy?
· What are your running costs?
· Are there any hidden expenses like necessary equipment? ie camera, films, slide projector, white boards, publications, tools, travelling expenses...?
· Do you have an emergency fund?
· Do you have a plan and the finance to carry it out?
· Do you intend to sell publications, plants, design consultancy, landscaping services or other forms of appropriate income generation to sustain you?

Costing

If you are running a business and submitting tax returns you will need to take setting up costs into account, which are all one-offs incurred when you first start up. Include everything you will need for work. This includes:
· Equipment.
· Tools.
· Stock (ie books, magazines, products, plants, tools for resale and for your research).

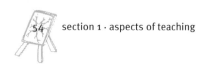

· Stationery.
· Account books.
· A filing system.
· A computer.

All these details add up to a considerable amount of money. People usually underestimate on start up costs and it is a wise move to add at least 30% to your figure to account for hidden expenses. Also your running costs will need to be listed. You have to pay these even if you produce, supply or sell nothing at all. These include rent, rates, light/heating, repairs/maintenance, telephone, insurance, vehicles, advertising and, of course, what you pay yourself.

When costing a course, it is useful to ask fellow teachers what they charge and whether they have discounts for students in full-time education, unemployed people, OAPs, and early booking. Do they charge proportionately by annual income? Remember that besides costing all the pre-course expenses and daily costs such as photocopying of handouts, travel, food etc, you must quantify all your time. We may never be adequately remunerated for all our permaculture work but we need to know what we are putting in for free. It is our gift and should be recognised by ourselves. Quantify it and then let it go. There is no point in resenting a gift and holding on to negative energy.

Self Assessment and accountants

Any self-employed business guide will give you the basics for keeping records for the Inland Revenue and it is a good idea to set your books up as soon as you start being self-employed.

The introduction of Self Assessment is, according to the Inland Revenue, a clearer and more straightforward way of working out and paying tax. The first Self Assessment tax returns were issued in April 1997 and covered the 1996-97 tax year. Called a 'major simplification' by the government, it makes the individual responsible for calculating tax and theoretically is meant to be saving the government money. We'll see!

All people in employment, whether a tax payer or not, will receive an eight page tax return with a step-by-step guide to on how to fill it in. There is also a guide on how to calculate your own tax, if you want to. If your annual turn over is below £15,000 for a full year, you may fill in the special, shortened income and expenses section. What this means is that unless you have other sources of income, it is unlikely that you will need an accountant to do this for you.

Expenses

When keeping accounts always maintain full and clear records of everything you buy and sell. Get receipts for everything. If you do have an accountant, he or she should tell you what you can claim as business expenses. This can include:

· Specialised clothing (ie for use with a chain saw).
· Tools.
· A proportion of car expenses which reflect business use.
· Travelling expenses.
· Fees paid to conference organisers and professional bodies.
· Heating, light and telephone expenses (calculated on how much space you use if you work from home).
· The costs of goods bought for resale and raw materials.
· Advertising.

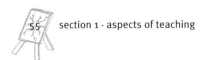

· Delivery charges.
· Cost of rented rooms.
· Postage and stationery.
· Design and printing costs.
· Relevant books and publications.
· Professional fees.
· Bank charges on business accounts.
· Wages.
· Training costs.
· Insurance premiums for the business.
· Overnight accommodation costs, dinner and breakfast (not lunch!) on trips away from home.

This is not a comprehensive list of working expenses – it is more of a prompt. There are now many publications and guides to becoming-self employed and running small businesses which are clear, comprehensive and are constantly updated. You can borrow them from a local library.

Marketing courses (and allied services)

Where do your permaculture students or design clients materialise from? No doubt your friends have heard all about permaculture and will be your first guinea pigs! They can become the core of a local group and will be of great assistance to you in the early days by spreading the word. Never underestimate the power of the grapevine. It is far more valuable than countless newspaper advertisements because of its personal nature. People are more willing to trust a good word in their ear than all the glossy ads in the world. They feel that they are less likely to be let down this way. Of course, one good word leads to another and your students will bring new people to you. The circle will expand and more and more people will hear of permaculture and consider taking one of your courses.

An important point to consider is that it can take years to establish your reputation by word of mouth. In the early days it may be necessary to teach courses to a very small audience and just break even with expenses. This way, at least you refine your course and teaching skills and get the word out about you. It may not support you financially but it is a start. The Permaculture Association (Britain) also run apprenticeship schemes for trainees, matching experienced teachers and practitioners with newcomers. This is a valuable way of acquiring skills and is highly recommended.

A fundamental support both for the teacher and for Design Course graduates is a local group. The local group can hold regular events like plant swaps, annual regional gatherings, workshops, slide shows and informal meetings. These activities attract new people to permaculture, create a situation in which resources like knowledge, plants, food and labour can be redistributed and provide support for permaculture people. They are immensely valuable. They are also fertile ground for potential students who can get to know you before committing themselves to a full course. The local permaculture group can be allied to the LETS group and other like-minded organisations. It can receive guidance from the Permaculture Association (Britain) and support (financial or otherwise) from the local council (either through their Environment or LA21 officer). If the Permaculture Association endorses the group and allows it to use their charitable status in funding bids, the Association also benefits from any funds raised. A valuable symbiosis is established. If you don't have a local group, start one! This is Zone 1 marketing, at your back door.

Getting known: free PR

Now all your friends and associates know of your work, what do you do? First, you need to explore all avenues of marketing that is free. Go to all the wholefood shops, environmental centres, libraries and natural health clinics in the area and introduce yourself. Ask them if you can put a small (well designed) flier on their noticeboard or leave some leaflets. Make friends. The best thing you can do, where appropriate, is be a customer as well. Practise what you preach.

Visit people like your local wildlife trusts, HDRA group or conservation group, introduce yourself and a leaflet. Offer to give talks, or better, slide shows on permaculture. There is nothing like a captive audience.

A trip to your local library to find out all the local organisations who have members who may be interested in permaculture is worthwhile. They will have the contact names and addresses of groups in the area. The main aim is to get your name known. You will be surprised how many contacts come to you through these free sources. You will be especially welcome if you can approach your subject from an interesting angle. Follow-up invitations will also come from other groups if you go down well.

Advertising

Take advantages of free listings in local papers, national and local permaculture publications, LETS mailings, the forthcoming events section of organisations like the Conservation Foundation, your local trusts, local council's publications, local Friends of the Earth and other eco groups etc. Again your local library or your local council will possibly hold this kind of data.

Newspaper advertising is expensive, even for classifieds. Make sure that your advertising copy is brief, to the point and interesting. If the first sentence catches the eye you are likely to receive enquiries. If you choose to advertise, always code every advertisement you place so that you can track whether it is cost effective. For example, place a sourcing code or number in your address or after your name. Say you are advertising in Permaculture Magazine you may want to use the issue number to source the responses. This lets you work out whether the cost of the ad is justified.

There is a logical line of thought that because the free papers are free, they are usually delivered unsolicited. As such, they often find their way post haste into the bin. However, the free papers in particular are usually hungry for news, any news, because generally they lack any! It is very likely that if you present your course or services as a news item, it will be snapped up with relish. Any publicity is worthwhile, however limited, especially if it is free.

When writing any leaflet, article or advertisement you need to attract the reader and hold their attention. The general direct marketing principle of 'AIDA' is a good one to follow: Attention, Interest, Desire, Action. The first thing you have to do is to grab the reader's Attention. This can be done using an appropriate headline and a good graphic – particularly on a poster or leaflet. Next, you need to create Interest by telling the reader more. When interest is achieved there needs to be a Desire to learn more. Finally, it is important to provide a need to Act, immediately preferably, as the longer it is left the less likely he or she will be to do anything. You can consider an incentive to book early like a discount with a cut off deadline to inspire action.

At all times, keep the reader's motivation to respond in mind by clearly defining the benefits of the course. Be very clear in presentation of contact details: your name, address and telephone plus a fax or email if you have them.

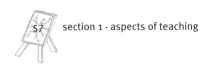

○ **Writing**

If you have a flair for writing, it is a fine way to establish your credentials as a permaculture teacher. You can write for local publications, newsletters, permaculture publications, general and specialised magazines and newspapers. Large newsagents are good for a browse and the library too will stock many publications, especially if you request them. You can then assess whether the publication you are interested in writing for is appropriate and will be interested in your subject. Research your subject well, whatever the scale of your creative endeavours, and spend time targeting your material.

Do not send unsolicited material to editors if you can have a chat first. But before that chat, proper research and consideration of the publication or newsletter you are going to talk to is essential. Who reads it? What angle is appropriate? What have they done before? Personal contact with an editor will also help you to appreciate what they need, style and content-wise. Appreciate that editors have serious pressures like deadlines, funding, circulation figures, the necessity of accommodating advertising to pay for the paper etc. Be professional, organised and meet your deadline. It will be appreciated. Remember illustrations are as important as the quality of the text. Provide them in abundance. Don't forget why you are writing the piece in the first place – to promote your courses or services. Work well in advance of your diary. Quarterly publications usually work at least half a year ahead of themselves. Take rejection with grace. It is a learning process which all writers have to experience. Be prepared to be edited. Everybody needs to be, even professional authors. Nobody's perfect.

Finally, have fun! Permaculture is exciting and transformative. Your love of it will be your most powerful ally.

○ **Bibliography**

Some parts of this article were extracted from Healthy Business – The Natural Practitioner's Guide to Success, Harland, M & Finn, G, Hyden House, 5th edition, 1998.

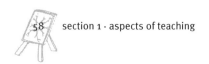

⇨ Example course formats

Two week residential · Joanne Tippett

Sunday

15.30 – 17.00	Arrival and registration
19.00	Get together

Monday

08.30 – 10.30	Introductions · Aims of the course · 'Problem Tree Analysis' – group work
11.00 – 12.30	Definition of permaculture · Introduction to mindmaps · Quality of life
13.30 – 15.00	Parable of the chicken · Structure of the course
15.30 – 17.00	The ecosphere: the science behind The Natural Step
19.00	Video: Global Gardener

Tuesday

08.30 – 10.30	What is sustainability? (based on The Natural Step) · Permaculture ethics
11.00 – 12.30	Ecological principles · connections, all waste = food, solar cascade
13.30 – 15.00	Walkabout · Introduction to design tools · Design tool: analysis of elements
15.30 – 17.00	Design tools: relative location and connective strategies · Zones
19.00	Slides of case studies/video

Wednesday

08.30 – 10.30	Ecological principles: diversity, succession, edge effect
11.00 – 12.30	The ecosphere: atmosphere and feedback loops, pedosphere and soils
13.30 – 15.00	Zone I and II practical eg Dorset garden and mulch
15.30 – 17.00	Planning guidelines: start small , build in feedback loops · Problem = solution · Minimum effort for maximum return · Work with nature
19.00	Topic: eg integrated pest management

Thursday

08.30 – 10.30	The ecosphere: hydrosphere, biosphere, and the role of trees
11.00 – 12.30	Design tools: sector analysis, stacking in space and time topic eg inter-cropping and agroforestry
13.30 – 15.00	Practical eg water harvesting, swales · Designing for water
15.30 – 17.00	Economics and community development · LETSystems
19.00	LETSystem game

Friday

08.30 – 10.30	Design tools: multiple uses for each element, multiple sources for each need · Local and biological resources
11.00 – 12.30	Design tool: designing for surge and pulse · Genetic and cultural diversity · Sustainability revisited
13.30 – 15.00	Design practical: small scale design exercise
15.30 – 17.00	Design tools: cycling of energy and materials · Appropriate technology
19.00	Topic: eg animal systems

Saturday

08.30 – 10.30	Design tool: pattern application
11.00 – 12.30	Review: group work
13.30 – 15.00	Practical: eg key hole bed
15.30 – 17.00	Summary of week – themes and patterns
19.00	Party

[Sunday]

Monday

08.30 – 10.30	Introduction to sustainable pattern design · Design practical: patterns in organisations, roles and guilds
11.00 – 12.30	Backcasting · Design practicals: quality of life · Values · Redefining roles and organisational structure
13.30 – 15.00	Design practicals: observation and resource inventory
15.30 – 17.00	Design practical: analysis of project themes in Design Process 1
19.00	Video: Global Gardener

Tuesday

08.30 – 10.30	Design practicals: goals and future economic elements
11.00 – 12.30	Design practicals: local resource inventory and limiting factors
13.30 – 15.00	Design practicals: decision making process for future economic elements and analysis of elements templates
15.30 – 17.00	Field trip
19.00	Topic – eg ecological architecture

Wednesday

08.30 – 10.30	Design practical: design information charts · Themes in Design Process 2
11.00 – 12.30	Design practicals: mapping and analysing the physical environment
13.30 – 15.00	Design practical: zones, sectors and elevation
15.30 – 17.00	Design practicals: water flow, water harvesting and bubble map
19.00	Free time or design continuation

Thursday

08.30 – 10.30	Pattern application · Design practical: rough sketch of the bubble map
11.00 – 12.30	Design practical: draw up the broad design
13.30 – 15.00	Design practical: detailed design (on the land)
15.30 – 17.00	Finish designs
19.00	Participants' work and presentations

Friday

08.30 – 10.30	Planning implementation and organisational structures
11.00 – 12.30	Design presentations and feedback
13.30 – 15.00	Summary of design process and applications
15.30 – 17.00	Networking and evaluation
19.00	Party

Breakfast 07.45 – 08.30 · Morning tea 10.30 – 11.00 · Lunch 12.30 – 13.30 · Afternoon tea 15.00 – 15.30 · Dinner 18.00 – 19.00

Two week residential Patrick Whitefield

Day one

09.30 – 11.00	Introductions · 'A tale of two chickens'
11.30 – 13.00	Discussion: ethics · Principles of permaculture I
14.00 – 15.30	Activity: site walk
16.00 – 17.30	Discussion: me and my home

Day two

09.30 – 11.00	Principles of permaculture II
11.30 – 13.00	Gardening I · Activity/discussion: what do I eat?
14.00 – 15.30	Activity: surveying skills (pacing, compass work and water level)
16.00 – 17.30	Slide show: an urban garden · Video: Global Gardener II

Day three

09.30 – 11.00	Case study: Jean Pain · Energy
11.30 – 13.00	Building · Activity: fruit (placement exercise)
14.00 – 15.30	Activity: practical work
16.00 – 17.30	Slide show: useful plants · Video: Global Gardener II

Day four

09.30 – 11.00	Gardening II · Slide show: mulching
11.30 – 13.00	Forest garden · Activity: communication skills (listening)
14.00 – 15.30	Visit to place of interest
16.00 – 17.30	

Day five

09.30 – 11.00	Soil · Slide show: farming
11.30 – 13.00	Grain · Discussion: agroforestry
14.00 – 15.30	Activity: soil practical
16.00 – 17.30	Slide show: reading the landscape · Video: Global Gardener III

Day six

09.30 – 11.00	Function of trees (brainstorm) · Spare
11.30 – 13.00	Short design exercise
14.00 – 15.30	
16.00 – 17.30	Communities

Day seven

09.30 – 11.00	Water · Slide show: water and sewage
11.30 – 13.00	Sewage treatment · Keylining (sandpit demonstration)
14.00 – 15.30	Activity: listening to the landscape
16.00 – 17.30	Spare · Video: Global Gardener VI

Day eight

09.30 – 11.00	Woodland I (windbreaks) · Slide show: trees and woodland
11.30 – 13.00	Design methods (including mapping)
14.00 – 15.30	Activity: base map exercise
16.00 – 17.30	

Day nine

09.30 – 11.00	Urban ecology · Woodland II (urban woodland)
11.30 – 13.00	Spare · Activity: design questionnaire
14.00 – 15.30	Activity: design questionnaire
16.00 – 17.30	Video and discussion: ancient futures

Day ten

09.30 – 11.00	Biodiversity
11.30 – 13.00	Community agriculture/direct selling · Spare
14.00 – 15.30	Activity: main design exercise
16.00 – 17.30	

Day eleven

09.30 – 11.00	Activity: main design exercise
11.30 – 13.00	
14.00 – 15.30	Activity: design presentations and debrief
16.00 – 17.30	

Day twelve

09.30 – 11.00	Biotime · Participants' presentation and requests
11.30 – 13.00	Participants' presentation and requests
14.00 – 15.30	Where do we go from here?
16.00 – 17.30	Feedback and close

Morning break 11.00 – 11.30 Lunch 13.00 – 14.00 Afternoon break 15.30 – 16.00

An alternative format for the Design Course which I have sometimes used is to split the 12 days into three blocks: an introductory weekend, followed by two five-day blocks. The introductory weekend is the standard Introductory Course (see page 63). People can come on this weekend alone, or they can take it as the first element of the Design Course. The two five-day blocks are centred over weekends, either Thursday to Monday, or Friday to Tuesday. There is usually a gap of two to three weeks between the blocks. The advantages of this format are:

· Combining the Introductory and Design courses makes each more viable in terms of student numbers.

· Some people who initially come only for the Introductory Course like it so much that they come back.

· People in regular employment often find it easier to take off a few days at the beginning or end of the week rather than a solid fortnight.

The disadvantage is the extra travelling involved, so this format is most suitable where a high proportion of the participants live locally.

Two week residential Nancy Woodhead and Graham Bell

The following curriculum represents a detailed listing of items included in a two-week course following the 72 hour course objectives. This can be altered at the start of the course to cover more or less of a particular element, depending on participants' interests and requirements. The final timetable is therefore discussed and agreed early in the course.

Introductions: Who is teaching · Where are we from · What do we expect/domestic notices

What is permaculture: An overview

What does it mean to be alive: Outdoor exercise

What fuels our hunger: Why do we need permaculture?

Pattern language: Structured thinking problem solving

The flywheel effect: Expectations and outcomes in the learning process

Everything is a gift: A group field exercise: Unused assets

Unused assets: Report back

Working as a team: Group strengthening exercises

Predators: Theoretical yield is unlimited

Soil: What is soil? Soil management 1

Real capital: 1. Homework

Real capital: 2. Resource management exercise – teams of three

Real capital: Report back

Money systems: What is money? What are the alternatives?

Soil: Letting soil do the work · Soil management 2

Where is your home?: Cultural inheritance/belonging to a place/ landscape

Home: Team exercise: what makes a home – built environment

Physical placement: Zones/sectors/edge

Stacking: The maximisation of time and space

Landscape: Let's look around/sandpit exercise

Mapping: Design through models/plans

Resource assessment: Individual exercise – mapping and listing this site

Wealth: The story of our community

Horticulture: A permaculture approach

Biodiversity: Individual exercise – species lists

A one day garden: (Practical) preparing ground/water

A one day garden: Stacking and niches · Planting out

Bioenergy: Trees and plant communities

Trees: Tree management and practice

Trees: Looking around us

Temperate landscape: Land form and management

Agriculture: Techniques for broadscale management

Smallholding: An integrative design exercise

Community agriculture: Alternative strategies

The design process: Methodology

Water: Observation and analysis: group exercise

Water: Management in theory

Aquaculture: Water as a medium for yield

Buildings and water: Design and implementation

Dwellings and workspace: Building design in general

Transport systems: Energy usage and management

Community design: Managing the process

Health in the landscape: Local inputs, herbs etc

Biotechnics: Geomancy for beginners

Group design: Practical project

Design checklist

Permaculture: Summarising portable indicators

Setting personal objectives: All

Review course: All

In addition the group are encouraged to create evening interaction, be it discussion, reading or entertainment. Bring your musical instrument, your story, your song. A concert on the last (or penultimate) evening where everyone performs something is a strong closing action.

All attendees will take an active part in the course and will be encouraged to present small or larger parts of the curriculum. They will be directed to tools for continuing learning. They will come away with an accurate inventory of their own and their communities' resources, tools to maximise their effectiveness, and ideas for regenerative work on resource-deprived aspects of their environment.

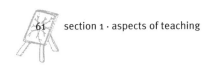

Twelve consecutive Saturdays Bryn Thomas

Day one – introduction

10.00 – 11.30	Welcome · Introductions · What do you want from the course?
11.45 – 13.15	Introduction to the course, domestics and administration · What is sustainability? · What is permaculture? · The parable of the chicken
14.15 – 15.45	Observation activity: Cedar Lodge Garden/Tivoli Copse · Group discussion (outdoors)
16.00 – 17.30	Patterns for learning · Permaculture ethics · Video: In grave danger of falling food

Day two – permaculture organisations, permaculture principles

10.00 – 11.30	Opening and announcements · Introducing design projects · The Pc community · The Pc association (Britain) · The Diploma in Pc Design
11.45 – 13.15	Principles 1
14.15 – 15.45	Activity: group work
16.00 – 17.30	Principles 2 · Design projects: identifying potential projects

Day three – design methods and process

10.00 – 11.30	Opening and announcements · Design projects · Forming design groups · Principles 3
11.45 – 13.15	Why design? · Design methods · Zone, sector and elevation planning
14.15 – 15.45	Design activity · Cedar Lodge · Zones and sectors
16.00 – 17.30	Design process · Designing for other people · Design projects: finalise group and projects

Day four – soil and water

10.00 – 11.30	Opening and announcements · Evaluation of the course so far · Observing soils · soil problems
11.45 – 13.15	Soil conservation and regeneration · Conserving soil and water in the landscape
14.15 – 15.45	Practical activities: making a swale, making a mulch bed
16.00 – 17.30	Visit to coast: coastal ecology and pollution

Day five – water

10.00 – 11.30	Opening and announcements · Characteristics of water · The water cycle and water uses · Water quality and pollution
11.45 – 13.15	Domestic water conservation · Rainwater harvesting · Sewage: waste or resource? · Treating and using · Compost toilets
14.15 – 15.45	Aquaculture · Mariculture · Activity: observing micro-climates
16.00 – 17.30	Trees and energy transactions · Modifying micro-climates · Gaia theory and climate change

Day six – farming and woodland

10.00 – 11.30	Opening and announcements · Fukuoka: do nothing · Farming · Sustainable cereals
11.45 – 13.15	Agroforestry · Fruit rootstocks and pollinators · Foggage pasture management · Leaf protein
14.15 – 15.45	Lunch in the woods? · Visit: Bonnie's wood · Woodland management and crafts · Wild foods
16.00 – 17.30	Woodland establishment · Wildlife and permaculture

Day seven – gardening and horticulture

10.00 – 11.30	Visit: Moulescombe Forest Garden
11.45 – 13.15	Opening and announcements · Permaculture in the garden · Urban gardening
14.15 – 15.45	Forest gardening · Edible landscaping · Garden exercise
16.00 – 17.30	Horticultural techniques · Plants and fungi · Case studies

Day eight – green buildings

10.00 – 11.30	Opening and announcements · Evaluation of course so far · Green architecture: an overview
11.45 – 13.15	Using the sun's heat · Building materials: energy and properties
14.15 – 15.45	Case study: future house, eco-renovation
16.00 – 17.30	Visit: 'Diggers' self-build houses, community self-build

Day nine – health, waste, energy, transport and bioregions

10.00 – 11.30	Opening and announcements · Bioregions · Case study: bioregional development group
11.45 – 13.15	Health and sustainability
14.15 – 15.45	Discussion groups: sustainable transport, renewable energy sources, energy conservation
16.00 – 17.30	Visit: Magpie recycling co-operative waste resources and recycling

Day ten – other climates and cultures; community and organisations

10.00 – 11.30	Opening and announcements · Arid and tropical environments · Problems, processes and strategies · Video excerpts: Global Gardener
11.45 – 13.15	'Third world' issues · Video excerpt: Ancient futures · Case study: Rural Nepal
14.15 – 15.45	Newspaper game · Agenda 21 · Tools for working with groups
16.00 – 17.30	Community business structures · Case study: Stanmer Organics · Housing co-operatives · Linking producers and consumers

Day eleven – money, economy and exchange

10.00 – 11.30	Opening and announcements · Real wealth and real capital · Economics and money · Sustainable exchange
11.45 – 13.15	Ethical investment · Credit unions and micro-credit · Funding projects
14.15 – 15.45	Local Exchange Trading Systems · Activity: LETS stimulation
16.00 – 17.30	Open space

Day twelve – design presentations

10.00 – 11.30	Opening and announcements · Design presentations
11.45 – 13.15	Design presentations
14.15 – 15.45	Design presentations · Networks and sources of information
16.00 – 17.30	Your ideas for the future · Evaluation

Morning break 11.30 – 11.45 · Lunch 13.15 – 14.15 · Afternoon break 15.45 – 16.00

Six weekends George Sobol and Patsy Garrard

Weekend one
Saturday
09.45 – 10.30	Introductions
11.00 – 11.45	Learning strategies
11.45 – 12.30	Introduction to permaculture
14.00 – 14.45	O'BREDIM and introduction to site
14.45 – 15.30	
16.00 – 16.45	Parable of the chicken
16.45 – 18.00	In grave danger...

Sunday
09.45 – 10.30	Ethics and principles
11.00 – 11.45	Introductions to each other
11.45 – 12.30	
14.00 – 14.45	A simple survey tool
14.45 – 15.30	Rainfall calculation
16.00 – 16.45	Guiding principles of permaculture
16.45 – 18.00	Close and clear up

Weekend two
Saturday
09.45 – 10.30	Pattern in nature
11.00 – 11.45	Zones and sectors
11.45 – 12.30	Base map theory
14.00 – 14.45	Mapping exercise
14.45 – 15.30	
16.00 – 16.45	Real wealth and wiser money
16.45 – 18.00	Permaculture in practice

Sunday
09.45 – 10.30	Pattern in design · Edge effect
11.00 – 11.45	Temperate landscape profile
11.45 – 12.30	Methods of permaculture design
14.00 – 14.45	Introduction to and survey with Bunyip A-Frame
14.45 – 15.30	
16.00 – 16.45	Global Gardener 1
16.45 – 18.00	Close and clear up

Weekend three
Saturday
09.45 – 10.30	Biological resources
11.00 – 11.45	Soils
11.45 – 12.30	No till systems
14.00 – 14.45	Soil and mulching practical
14.45 – 15.30	
16.00 – 16.45	Zone 1
16.45 – 18.00	Arid landscape

Sunday
09.45 – 10.30	Energy transactions of trees
11.00 – 11.45	Forest garden
11.45 – 12.30	Web of life · Zone 5
14.00 – 14.45	Practical: urban permaculture
14.45 – 15.30	
16.00 – 16.45	Global gardener 2
16.45 – 18.00	Close and clear up

Weekend four
Saturday
09.45 – 10.30	Living water
11.00 – 11.45	Sewage
11.45 – 12.30	Livestock · Zone 3
14.00 – 14.45	Practical: appropriate technology
14.45 – 15.30	
16.00 – 16.45	Zone 4 · Woodland management
16.45 – 18.00	

Sunday
09.45 – 10.30	Aquaculture
11.00 – 11.45	New build
11.45 – 12.30	Bioregions
14.00 – 14.45	Practical: inputs/outputs of the home
14.45 – 15.30	
16.00 – 16.45	Global gardener 3
16.45 – 18.00	Close and clear up

Weekend five
Saturday
09.45 – 10.30	Presentation of a design
11.00 – 11.45	Client interview
11.45 – 12.30	
14.00 – 14.45	Work in groups – site survey and record
14.45 – 15.30	
16.00 – 16.45	Record and evaluate
16.45 – 18.00	Global gardener 4

Sunday
09.45 – 10.30	Drawing skills
11.00 – 11.45	Further questions · design work in groups
11.45 – 12.30	
14.00 – 14.45	Design work in groups
14.45 – 15.30	
16.00 – 16.45	Design groups feedback and planning
16.45 – 18.00	Close and clear up

Weekend six
Saturday
09.45 – 10.30	Design groups meet
11.00 – 11.45	Presentations
11.45 – 12.30	
14.00 – 14.45	Presentations
14.45 – 15.30	
16.00 – 16.45	Presentations
16.45 – 18.00	

Sunday
09.45 – 10.30	Development strategies
11.00 – 11.45	Working groups that work
11.45 – 12.30	
14.00 – 14.45	
14.45 – 15.30	Where do we go from here?
16.00 – 16.45	Course feedback
16.45 – 18.00	Close and clear up

Opening and introductions 09.00 – 09.45 · Tea 10.30 – 11.00 · Lunch 12.30 – 14.00 · Tea 15.30 – 16.00 · Supper 18.00

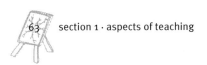
Introductory courses

Simon Pratt

Friday

19.00 Dinner
20.00 Introduction to Redfield
20.30 Bill Mollison's video: In grave danger of falling food (or Global gardener)

Saturday

08.30 Breakfast
09.30 Introductions
10.15 Parable of the chicken
10.45 Break
11.15 Ethics and principles of permaculture
11.45 Tour of the site
13.00 Lunch
14.15 Zones, sectors, elevation planning
15.00 Minimum intervention, edge, stacking
15.45 Break
16.15 Permaculture in practice: Mulched garden
16.45 Practical work
18.30 Dinner
20.00 Social event(s)

Sunday

08.30 Breakfast
09.30 Review
10.00 Alternative financial systems
10.45 Break
11.15 Design sequence
11.45 Practical design work
13.00 Lunch
14.15 Design work and presentations
15.30 Where do we go from here?
16.00 Evaluation and leave taking
16.30 Close

Patrick Whitefield

Saturday

09.30 Introductions
10.15 A tale of two chickens
11.00 Break
11.30 Principles of Permaculture
13.00 Lunch
14.00 Walk
14.45 Forest garden
15.30 Tea
16.00 Video: In grave danger of falling food
17.00 Close

Sunday

09.30 LETS and community agriculture
10.15 Permaculture gardening
11.00 Break
11.30 The design process
12.15 Designing (on site)
13.00 Lunch
14.00 Design (on paper)
14.45 Design presentations
15.30 Tea
16.00 Slide show: Practical mulching
16.30 Feedback and Where do we go from here?
17.00 Close

Central themes

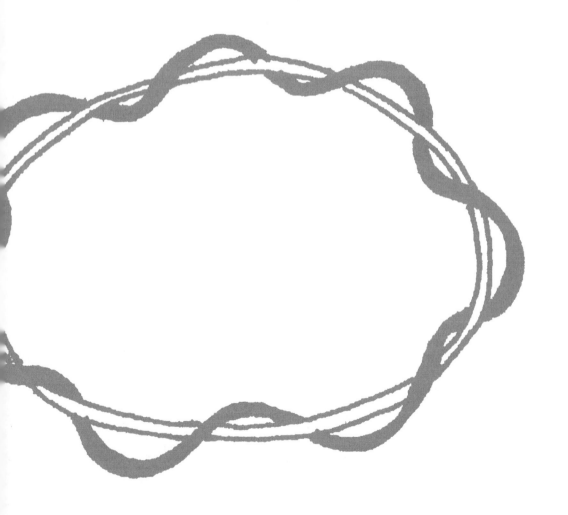

Central Themes

The central themes of permaculture are usually considered to be Ethics, Principles and Design. However, during the preparation of this book, it became increasingly clear that there is another – Observation. It is fundamental to the permaculture approach, and so we have included it here as a central theme.

An awareness and understanding of Agenda 21 is important for any graduate wishing to be involved in wider community change. It is central to many of the projects and activities which permaculture designers are involved in, and so it has also been included here as a central theme.

The sessions within this section have been put into five chapters:

Permaculture Design is unusual in that it contains and is built upon an agreed ethical framework. This makes it radically different from subjects such as organic gardening, alternative technology or agroforestry where the ethos is implicit rather than explicit. Teaching the ethics provides one obvious way of drawing a distinction between Permaculture Design and other systems.

In this chapter, Chris Dixon and Patrick Whitefield present how they teach the ethics of permaculture.

Broad objectives for ethics sessions:

· To present the three permaculture ethics.
· To explain that permaculture shares these ethics with many people around the world.

The ethics as defined in Permaculture, A Designers' Manual, by Bill Mollison:

The Ethical Basis of Permaculture

1 Care of the earth:

Provision for all life systems to continue and multiply.

2 Care of people:

Provision for people to access those resources necessary to their existence.

3 Setting limits to population and consumption:

By governing our own needs, we can set resources aside to further the above principles.

Additionally

The Prime Directive of Permaculture

The only ethical decision is to take responsibility for our own existence and that of our children.

 # Ethics of Permaculture Design

Chris Dixon

○ **Objective**

To generate thinking which is inclusive and holistic rather than fragmentary, through the use of a simple framework that recognises inter-penetration and complexity.

○ **Learning outcomes**

By the end of this session students will be able to:

· State each ethic and give simple examples.
· Explain the ethics in terms of a pattern towards holistic thinking.
· Demonstrate the inter-penetration of ethics.
· Explain how the ethics can be used as a guide to design work.

○ **Context**

I use the ethics as an introduction to the course. As they require examples, they can be used to suggest something of the inclusiveness of the subject.

○ **Duration**

30 minutes

○ **How I teach this session**

Interactive talk

30 minutes

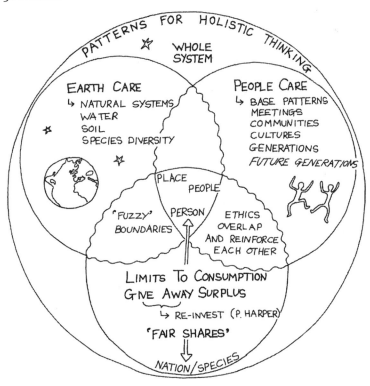

I introduce the ethics of permaculture design as a Pattern for Holistic Thinking. I generate the pattern over the course of the session on the board, drawing as much of the information and thinking as possible from the group (see above). I begin with the large circle which I call the whole system and then ask for the ethics and include them in their own separate but overlapping circles. I describe the boundaries between them as fuzzy (inter-penetration). I ask for examples relating to each ethic and include them if the group agrees they are relevant. I also make sure the ones I consider of particular importance get in there.

Additional notes
· By being able to include the word Person in the inner group, along with Place and People, I can present the ethics as a very powerful design tool which provides us with the three main areas of attention required in effective, sustainable design.
· I use the version of the third ethic that appears in the Designers' Manual (Ref 1) ie Setting Limits to Population and Consumption. I find this particularly useful as it can be used both at a personal level (as a permaculture designer, I choose these limits for myself), or at a national level (as a country, we choose not to use land mines etc) or even at a species level.

· I also mention modifications to the third ethic. For example, Peter Harper suggests 'reinvest surplus' rather than 'give away' (Ref 2). Similarly, in the early 1990s, the keyword Fairshares was coined as a replacement description, by Danish architect Tony Anderson (Ref 3). I describe a number of complications and possible challenges that may arise when using this latter term.

○ Link to principles

I find it convenient to move from the ethics to an introduction to the principles. Practical examples of principles can be related back to the ethics.

○ Further activities

· Successive subjects can be related back to the ethics during the course, providing further understanding and applications.

· A longer session of say one hour can be included part way through the course to review their use and application.

· At the end of the course, students can also be asked to swap time in pairs, two minutes each way, on the question of how the course itself measured up to the ethics. The results can be fed back and collated during final circles as part of feedback.

○ Bibliography

1 Permaculture: A Designers' Manual
2&3 Reinvest surplus and Fairshares have emerged within informal contexts, written references are not available

Discussing ethics

I find that most people are comfortable with the idea of 'ground rules' or 'rules of the game'. By explaining permaculture as a 'game', participants can make up their own mind about whether they become 'players' or not. I think that it's counter-productive to try to require people to behave ethically, no matter what the situation that they find themselves in. But to ask for a certain approach if people are involved in permaculture, then that's fine. We can't force people to play the game, but we can be clear about what the game is, and how we play it. The choice to participate remains with the individual.

The dynamic of every group I teach is different. I am interested in the participants being comfortable early on in the course. Challenges can come later when the participants feel safer with each other. I like to get a sense of the group's cultural norms rather than immediately imposing my own.

As an example, dealing with Earth Care is not always an easy option in neighbourhoods where there is more concrete than earth. I try to start with people where they are. If social issues are in the forefront, then I start there.

Limits to Population and Consumption can also be a challenging area to open up. The question of both distribution and limits can put up mental barriers in some people, with the assumption that someone is going to further limit their available resources, which may appear, to them at any rate, to be pitiful by today's standards. The message I try to get across is that we are going to limit our own resource use, because it is in our interest to do so. Most objectors will then relax about this.

Some groups respond well to stories of voluntary starvation by tribal peoples in order to preserve the species that the tribe depends upon, whereas others find this difficult to relate to.

What does work for me is when I stick to the two principles, necessitous use (use something only if you have to) and conservative use (if you use something, put it back) – the story of the oak beams used to repair New College Oxford from oak planted for this purpose when the college was built, retold recently on the TV programme 'How Buildings Learn', usually gets the point across.

Angus Soutar

Bibliography
1 How Buildings Learn, Brand, S.

Further research
"From Reductionism to Holism in Ecology and Deep Ecology", Rowe J S, The Ecologist, Vol 27, No 4, July/August 1997.

Ethics of Permaculture Design

Patrick Whitefield

○ Objective

To present the three permaculture ethics, and to give students an opportunity to reflect on and discuss their own ethical stance.

○ Learning outcomes

By the end of this session students will be able to:
· State the permaculture ethics.
· Understand them.

They will have had the opportunity to:
· Ask the question 'What is my own ethical stance?'
· Compare their own ethics with the permaculture ethics and those expressed by other students in discussion.
· Bring up specific ethical issues for general discussion.

○ Context

I teach this topic on the first day, after introducing permaculture using A Tale of Two Chickens as an example. (See pages 183-186.) The ethics are taught before going into the principles in detail. Since this is mainly a discussion session, it balances the more informative sessions on either side of it.

○ Duration

45 minutes

○ How I teach this session

Talk

10 minutes

I draw a triangle on the board - Earth Care, People Care, Limits and say a few words about each.

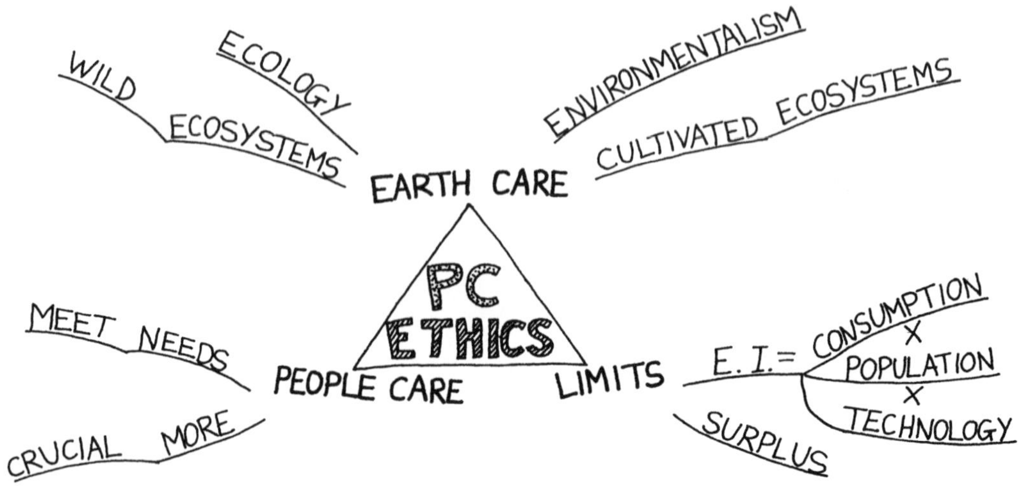

Earth Care

· Note the difference between environmentalism and deep ecology:

  Environmentalism sees humans as separate from nature, and talks about resources.

  Deep ecology sees us as part of nature, and talks about communities.

· Both are valid, and are views which can be held by the same person at the same time.

· Preservation of wild ecosystems: greening of human systems.

People Care

· Designs must meet people's needs or they either won't be implemented or will oppress people.

· Our interpersonal skills are less developed than our technical, Earth Care skills; most big problems are emotional and socio-political ones, not technical ones.

Limits to Population and Consumption

· The idea that we need to live within the Earth's physical limits is a liberating one. Analogy with children given loving limits compared to those given none.

· Level of consumption more crucial than size of population (Ref 1).

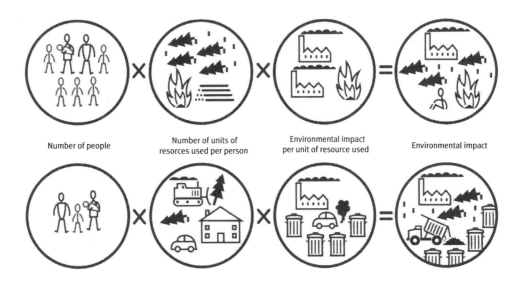

| Number of people | Number of units of resorces used per person | Environmental impact per unit of resource used | Environmental impact |

Ecological impact = population x level of consumption x technology

Small group discussion

10-15 minutes

· Form groups of three for discussion. Topic: Do you have a personal set of ethics? How do they compare with the ones I've just outlined?'

Whole group discussion

Time variable

· Has anyone got anything arising from that discussion they would like to say to the whole group?' General discussion may ensue.

○ Bibliography

1 Living in the Environment

In this chapter, three different approaches are illustrated, and show how the principles can be covered in two 45 minute sessions (1), three 90 minute sessions (2), or used to structure the entire first half of a course (3). The latter includes background information on the principles of the Natural Step.

Spending different amounts of time on the principles, clearly affects the timing and structure of the rest of the course, and the timetables on pages 58-63 show this.

'Cycling' gives an example of a detailed session on one of the principles, whilst 'Principles in Pictures', and the 'Principles Review Exercise', are both sessions which help to increase the student's ability to remember and understand the principles.

Broad objectives for principles sessions:
· To present the principles of permaculture.
· To provide examples of their application.
· To show how they have been derived from a combination of observation of natural systems, and the work of different practitioner's experience.
· To create a framework of understanding which is developed through the course.

Few permaculture teachers use exactly the same set of principles, although many of these differences are ones of wording. Many devise their own lists, which are made up from a combination of their own observations and thinking, and the two sets of principles given in the following Australian permaculture books:
From Permaculture, A Designers' Manual, by Bill Mollison:
· Work with nature rather than against.
· The problem is the solution.
· Make the least change for the greatest possible effect.
· The yield of a system is theoretically unlimited.
· Everything gardens.
From Introduction to Permaculture, by Bill Mollison & Reny Mia Slay:
· Relative location.
· Each element performs many functions.
· Each important function is supported by many elements.
· Efficient energy planning: zone, sector and slope.
· Using biological resources.
· Cycling of energy, nutrients, resources.
· Small-scale intensive systems; including plant stacking and time stacking.
· Accelerating succession and evolution.
· Diversity; including guilds.
· Edge effects.
· Attitudinal principles: everything works both ways, and permaculture is information and imagination-intensive.

A recent search on the Internet found a site which listed over 50 'permaculture principles'. This proliferation of principles may well be an area which needs to be addressed, but the editor's felt this to be beyond the remit of this project.

Permaculture Principles

Patrick Whitefield

O Objective

To present the principles of permaculture and give examples of them. Thus to give an overview of permaculture and the framework which is fleshed out by the rest of the course contents.

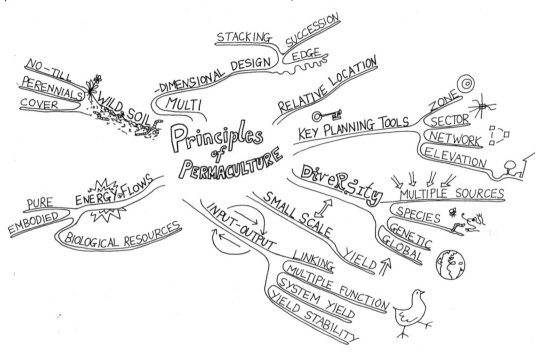

O Learning outcomes

By the end of this session students will be able to:

· State the principles of permaculture.
· Understand them.
· Have an idea of how these can be relevant or useful to them personally.
· Have a broad overview of permaculture.

O Context

Either on Day 1 or split between Days 1 and 2. Preceding it are: an Introduction to Permaculture - by way of an example, the chicken; and the session on Ethics. (See my Example Course Format on page 59.)

Sustainable Design

The concept of sustainable design is fundamental to my teaching of permaculture. It is relevant to Ethics, Principles and Design Methods, and I usually introduce it when it arises in discussion rather than at any fixed point. If it has not arisen by the Design Methods session, I start that session off with it. I sketch two concepts of design (see diagram 74). The first represents the conventional idea of design, an active process, with the designer as the major influence on the final design. The second represents the permaculture or sustainable approach to design, a more passive, receptive process, with the designer as listener, and the land and the people as the two major influences on the design.

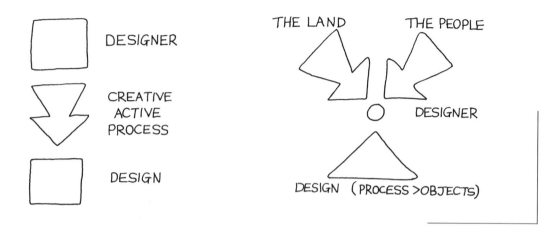

○ Duration

90-120 minutes

○ How I teach this session

I combine talk, slides and discussion. In order not to have too long a time in any one of these, I divide the principles into two groups.

For a complete list of the principles as I teach them, see the mind-map (previous page). I have evolved this set of principles from a combination of: the ideas set out in Permaculture One, the list of principles in the Introduction to Permaculture, David Holmgren's Hepburn Permaculture Gardens, and my own thinking (Refs 1, 2 & 3).

The order in which I take the principles reflects the historical development of permaculture ideas: starting with those which most closely imitate natural ecosystems, and going on to those which learn from the principles of how ecosystems work rather than copying them directly.

Part 1: Talk

30 minutes

a) The ecosystem as model for permaculture, see pp2-3 of Permaculture in a Nutshell (Ref 4)

b) Introduce the following principles:

Wild soil

· No-till.

· Perennial plants.

· Covered soil.

I point out that this is not a dogma but a preference: tilling and annual plants can have a place in permaculture, but given the choice we prefer undisturbed soil and perennial plants. The same goes for all the principles.

Multi-dimensional design

· Stacking.

· Succession.

· Edge: especially the woodland edge and that between water, land and air.

I point out how these three interact, especially in a temperate climate.

Relative Location

Return to the ecosystem as model: what makes an ecosystem work is the web of beneficial relationships between its components, therefore where we put things relative to each other is of the essence, this is why permaculture is so much about design. Eg the chicken-greenhouse, the importance of growing food in towns.

Edible Ecosystems ⋯⟩ Beneficial Relationships ⋯⟩ Relative Location ⋯⟩ Design

Natural ecosystems are the inspiration for permaculture. Our aim is to create systems which provide for our needs and yet have the low input requirement, low pollution output and sustainability of natural ecosystems. We can call these Edible Ecosystems, although of course they supply us with much else besides food.

Sometimes permaculture systems look like natural ecosystems, but more often they imitate the principles on which ecosystems work rather than their outward appearance. The most fundamental of these principles is the network of Beneficial Relationships between the different components.

If we're trying to make Beneficial Relationships between the components of our systems, where we put things in relation to each other is obviously of the essence. This is why Relative Location is so important, and why permaculture is above all a Design system.

Part 2: Slide show

10 minutes

Ten slides illustrating the above principles, including:

· No-dig gardening.
· Stacking in Zone 3, Zone 2 and Zone 0.
· Relative location: an example of plant-plant interaction in nature, and an example of the same principle used to guide plant location in an urban garden.

Part 3: Discussion

15 minutes

· In groups of three, discuss the topic, How can I make use in my own life of the principles mentioned so far?

Part 4: Talk

20 minutes

Introduce the following principles:

Key planning tools

· Zone: give the principle and also list the conventional zones 0-5.
· Sector: give the principle, list the factors involved, and introduce the concept of micro-climate.
· Network.
· Elevation: draw the classical humid landscape profile, and point out the advantages of working with the landscape rather than against.
· Integration: note that these four combined make a powerful planning system.

Network analysis

Like zone and sector, network is a concept borrowed from geography. While zone analysis looks at relationships with the main centre of human activity on a site, network looks at relationships between a number of different centres of activity within the site.

I use the example of David Holmgren's homestead (see Hepburn Permaculture Gardens, pp9 & 31). In designing the smallholding he had to consider flows of people, animals, materials and vehicles between: house, barn, a potential second house site, gardens, orchard and pond.

A simpler example would be the relationship between: compost bins, comfrey patch, kitchen door, main vegetable beds.

Diversity

· Multiple sources: on a global and a home garden scale.
· Species diversity.
· Genetic diversity.
· Global biodiversity: note its relationship with permaculture.

Small scale

· Its relationship with diversity.
· The relationship between scale and yield.

Input-output

· Linking up outputs and inputs within a system: refer back to the chicken.
· Multiple output, in terms of:
  · individual components within a system
  · the system as a whole
  · widening the concept of yield
· Yield stability: sustainability is the aim, not short-term maximum yield.

Energy flows

· Energy in use, a one-way flow.
· Embodied energy, cyclical flow is possible.
· Biological resources.

Part 5: Slide show

10 minutes

Eleven slides illustrating:

· Zone, sector and elevation on both domestic and broad scales.
· The relationship between scale and diversity.
· Multiple outputs.
· Pig-greenhouse.
· Biological resources: chicken tractor, slug predators.

Part 6: Discussion

15 minutes

· In groups of three, discuss the topic, How can I make use in my own life of the ideas mentioned in the latter part of this session?

NB: If all the principles are covered on the morning of the first day, this discussion comes just before lunch at the end of a long, intensive session. In that case I drop it, because the energy for discussion is never there.

○ **Bibliography**

1 Permaculture One
2 Introduction to Permaculture
3 Hepburn Permaculture Gardens
4 Permaculture in a Nutshell

○ **Further research**

Wild soil:

Permaculture in a Nutshell. Garden scale, pp26-34. Farm scale, pp42-47

Designing & Maintaining Your Edible Landscape Naturally. Garden scale, pp95-111

The One-Straw Revolution. A classic of no-till farming

Multi-dimensional design:

How to Make a Forest Garden

Key planning tools:

Permaculture in a Nutshell, pp23-25 & 37-39

Hepburn Permaculture Gardens

Diversity:

Agroecology, the science of sustainable agriculture, Altieri, M, 2nd edn, Intermediate Technology Publications, 1995. Especially for species diversity in agroecosystems.

Saving the Seed, Vellvé, R, Earthscan, 1992. Genetic diversity in crop plants.

Living in the Environment. Global biodiversity.

The Diversity of Life, Wilson, E O, Allen Lane, 1992. Global biodiversity.

Small scale:

The Garden Controversy, a critical analysis of the evidence and arguments relating to the production of food from gardens and farmland, Best, R H, & Ward, J T, Dept of Agricultural Economics, Wye College, Kent, 1956.

Smallholders, Householders, farm families and the ecology of intensive, sustainable agriculture, Netting, R M, Stanford University Press, 1993

Energy flows:

'Energy and Permaculture', Holmgren, D, in The Permaculture Edge, Vol 3, No 3, 1993.

Entropy, a New World View, Rifkin, J, Paladin, 1985.

Principles of Permaculture Design

Bryn Thomas

○ Objective

To provide course students with an overview of permaculture principles and some examples of their applications.

○ Learning outcomes

By the end of these sessions students will be able to:
· Understand, explain and illustrate the principles of permaculture design.
· Begin to apply these principles in design, understanding that it is the integrated use of most if not all of these principles that is necessary for effective sustainable design.
· Take part in group discussions and activities with greater confidence.
· Identify a number of problems that need to be addressed using sustainable design.
· Continue with the rest of the course.

○ Context

The rest of the course builds on the Principles. I normally cover them on Days 2 and 3. They follow Ethics on Day 1 and lead into Design Methods and Processes. I try to split up the principles sessions; it's all too much in one lump.

I also like to precede the sessions on principles with a 90 minute observation activity and discussion, preferably in a semi-natural habitat. Woodland is ideal. This focuses on our innate ability to make useful observations and starts to draw out permaculture principles from them. A site walk and discussion can also usefully follow the sessions on principles. (See my Example Course Format on (page 61.)

○ Duration

Three 90 minute sessions

○ How I teach these sessions

I think all teachers find structuring sessions on the principles hard work. What are they? Various texts list them differently, omitting some, giving others different names and presenting some as sub-divisions of others. This may well be confusing to students, and they should be warned about these inconsistencies. But all classifications are false, and looking at them differently broadens understanding.

Nevertheless, we need a structure, and I use the one below , to add to the confusion! Grouping them in threes gives an opportunity for a short break in the middle of each session.

Each principle is introduced with examples and where possible I use an activity or slides. The examples and practical activities used in the rest of the course help to broaden the students' understanding of the principles.

Session 1

Overall introduction to principles
Talk

5 minutes

Working With Nature (40 minutes)

1 Use patterns
Interactive talk

10 minutes

The concept is simple and requires little explanation.

· Ecological and natural patterns, eg river forms and flow, and disastrous
  consequences of intervention.
· Cultural patterns and wisdom, eg biotime and planting calendars. (See 'Observation'
  Simon Pratt, pages 144-145.)
· Summary:
   · Necessity for observation. The role of science.
   · All of the permaculture principles are patterns.

Slides or acetates

10 minutes

Examples and applications of natural patterns.

2 Minimum intervention/Work With Not Against Nature/Least Change For Greatest Effect
Interactive talk

10 minutes

Opposite to spirals of destruction, where an initial poorly conceived intervention causes the need for
ever increasing interventions.

· Hierarchy of intervention:
   · Do nothing
   · Biological Intervention
   · Physical Intervention   [interchangeable in
   · Chemical Intervention   certain situations]
· Work where it counts, eg Chris Dixon's approach to smallholding in West Wales. (See 'Regeneration'
  by Chris Dixon, pages 204-207.)

3 Perennial and no-dig systems
Interactive talk

10 minutes

· Everything gardens.
· Perennial vegetation requires less maintenance.
· Undisturbed soil maintains and cycles fertility most effectively.
· Example: wheat field versus woodland.

Diversity (40 minutes)

4 Diversity
Interactive talk

10 minutes

· Diversity = stability.
· Balanced systems contain producers, consumers and decomposers.

- Polycultures – sum of yields is greater than monocultures, eg: beans, maize and pumpkins; traditional orchard.
- Monocultures are maintained disorder – they require an input of work/energy to maintain, eg wheat; industrial towns dependent on a single industry.

Activity: Web of life

15 minutes

Give everyone a card representing an element in a system, eg a garden with a tree, a pond, a gardener, a compost heap, a kitchen, comfrey, chickens, etc. Stand everyone in a circle and ask each of them to think about which other elements he or she may have connections with. Ask one of them to throw a ball of cotton, whilst holding on to the end, to another element to which they consider they might be connected. Continue to form a strong and resilient web. Now disaster: eg chop down the tree. The person holding this element lets go. Observe the knock on effects as this and other elements die and let go.

5 Relative location/Maximise beneficial relationships/Integration

Interactive talk

10 minutes

- Diversity of useful connections not diversity of elements. Results in stable systems and high yields. Don't force relationships.
- Guilds: light/shade, water, nutrients, root zones, eg: maize, beans and squashes; plum, blackcurrant, comfrey.
- Maximise co-operation, minimise competition, eg: permaculture chicken; kitchen and vegetable garden. (See Designers' Manual pp24-25, Ref 1.)

6 Niche/Optimum location

Interactive talk

5 minutes

- Explain concept, eg blackcurrant requires moisture, fertility and shelter; figs require sunny sheltered spots with thin soil to fruit well; human beings are more productive when healthy and content.

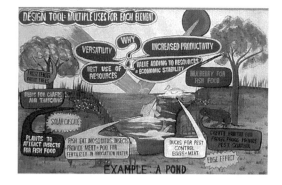

Session 2

Function and scale (40 minutes)

7 Each element should perform many functions

Interactive talk

8 minutes

- Ideally three or more functions, eg: permaculture chicken; shelter belt (shelter, fuel, timber, fruit/nuts, bee fodder, animal fodder, etc); living room.
- Select multi-functional elements, eg: lime tree (timber, fuel, charcoal, bee fodder, animal fodder, flowers for tea, etc); sea buckthorn (pioneer, exposure tolerant, fast growing, nitrogen fixing, vitamin C rich fruit).
- Multi-function dependent on relative location.

8 Important functions should be supported by many elements

Interactive talk

8 minutes

- Most necessary for: food, water, energy.
- Negative example: Irish potato famine.
- Fact: 75% of the worlds calories are provided by rice, maize, wheat and potatoes.

9 Small scale intensive systems/Optimum sizing

Interactive talk

8 minutes

- Modern agriculture is input and capital intensive, consequently it is not efficient despite high yields per unit area. It can therefore be viewed as extensive, despite being called intensive. This also applies to many other contemporary systems.
- Permaculture systems, once established, should require few external inputs and be higher yielding due to diversity and multi-function. They are truly intensive.
- Many things have an optimum size, eg a flock of one chicken is too small, a battery shed full is too big; dam sizes.
- This frees up unused land to 'wilderness'.

Slide show

16 minutes

I use various slides at the end of the session to illustrate principles 7-9.

Use of space and time (40 minutes)

10 Stacking

Interactive talk

8 minutes

- Optimum use of three dimensional space.

Video excerpt

7 minutes

Global Gardener or Forest Gardening – Robert Hart describing levels of a forest (Refs 2 & 3).

11 Use edge effect (ecotone)

Interactive talk

8 minutes

- Maximise in most circumstances, eg woodland; pond; estuary; trees as edge between ground and air.
- Contain species from both climax and edge species, so greatest diversity.
- Edges accumulate nutrients and organic matter; exchange nutrients, matter and information.
- Edges to increase: ponds; woodlands.
- Example design features: graded edges; sun facing arks; herb spiral.
- Humans are an edge species.
- Edges not to increase: surface area of a dwelling – this causes heat loss; edges between beds and weeds.

12 Use, mimic or accelerate natural succession and seasonal change

Interactive talk

8 minutes

- Succession, eg open ground to woodland; pond to woodland.
- Diversity and productivity increase with succession, then tail off a bit.

- Changes:
  - Initial stages are competitive and wasteful in a harsh environment.
  - Climax stage is more co-operative with more efficient cycling.
- Rolling permaculture.
- Plant pioneer and climax species together.
- Analogy to growth of the permaculture movement.
- Encourage students to consider their role.
- Seasonal changes: niche in time/stacking in time/stacking in the fourth dimension.
- Example: plant growth in woodlands through a season, bulbs in January, top trees in May.

Slide show

10-15 minutes

I use various slides at the end of the session to illustrate principles 10-12.

Session 3

Resources (20 minutes)

13 Use biological and renewable resources

Interactive talk

3 minutes

- Harvest only at sustainable levels.
- Avoid non-renewable resources where possible.
- Avoid the use of resources that pollute.

14 Use appropriate technology

Interactive talk

3 minutes

15 Stability/Self-regulation/Homoeostasis/Feedback

Interactive talk

4 minutes

- Self-reproducing systems.
- Establishment of carrying capacity.
- Plan for disasters/build in resilience.
- Dynamic equilibrium.

Slide show

10 minutes

Energy (20 minutes)

16 Energy efficient planning

Interactive talk

3 minutes

- Start close, start small.
- I refer to zone sector and elevation planning here but do not cover it until Design Methods and Processes.

17 Energy and material cycling/Maximise the use of energy from source to sink

Interactive talk

7 minutes

- Examples: life; water; soil; heat; money.

- Food chain/web of life: trophic pyramid, vegetarian v. meat issues. Caution – certain members of your group will almost certainly want to debate this for at least an hour. I recommend you acknowledge this as an important ethical issue, but emphasize the inclusive approach of permaculture and limited time availability.
- Reduce, reuse, recycle, resource origin.
- Design to conserve or replace an equal or greater resource than used. A system should produce more energy than it consumes.

18 Establish low maintenance systems / 80:20 principle

Interactive talk

10 minutes

- Unmet needs = work.
- Unused output = pollution.
- Many modern systems = maintained disorder = energy input.
- Stable systems (mostly natural) = ordered, but may appear chaotic = energy neutral or available yield.
- Example: around 10 calories of energy input required to provide one calorie of food eaten. This is an average for all foods and includes meat which is even more inefficient than 10:1.

Activity

5-10 minutes

Split into groups of three or four. Give each group two opposing situations (picture cards?), eg wheat field v. semi-natural woodland, canalised v. meandering river, grid v. evolved street plan. Ask them which is ordered and which is chaotic, and why? Discuss in groups for three minutes and feedback to the whole group for two minutes.

Attitudinal principles (30-50 minutes)

19 Everything works both ways/Problems contain their solutions/Everything is a gift

Discussion

5 minutes

- Does this always hold true? Nuclear waste?

Activity

10 minutes

Split into groups of two, three or four people. Encourage each person to think of a practical problem and how viewing it differently can present the solution (6 minutes). Close with a circle briefly sharing what was discussed (4 minutes).

20 Permaculture is information and imagination intensive

Interactive talk

5 minutes

- Make mistakes on paper.
- Yield is theoretically unlimited.

Activity

10 minutes

Present very simple example designs on paper to groups of three or four students and ask them to identify additional yields. For example fungi and recreation could be additional yields for a forest garden (6 minutes). Feedback each group's ideas to the whole class (4 minutes). Each group can have a different design.

21 Design to minimise the effects of limiting factors

Interactive talk

10 minutes

· Example: growth rate of grass and herbs in pasture over a year, represented graphically:
    · Winter: limiting factor to plant growth is temperature, so increase shelter with trees.
    · Mid summer: limiting factor to plant growth is availability of water, so increase shelter to reduce water loss, and increase soil water storage by increasing absorption (swales, organic matter, structure).
    · Spring: limiting factor to plant growth is nutrient availability, so improve fertility.

Whole group activity

10 minutes

Brainstorm some general limiting factors under two headings – physical, eg: climate, soil; and invisible, eg money, beliefs.

Activity in pairs

10 minutes

Ask students to identify something they want to achieve and what limits them in achieving this. In pairs ask them to look at the limiting factor 'with a sideways glance'. I have known this activity to bring up emotional stuff for people who have wanted to blame others for their problems and realised that they are responsible. For this reason I would not recommend feeding back from pairs to the whole group.

Slide show

10-15 minutes

I use various slides at the end of the session to illustrate principles 13-21.

Link work (for part time courses).

Recap the principles covered and consider how they relate to your life now. What changes can be made in the future? Read any of the references available.

○ Bibliography

1 Permaculture: A Designers' Manual
2 Global gardener
3 Forest Gardening

○ Further research

Introduction to Permaculture. Highly recommended as main reference.

Earth User's Guide to Permaculture. Recommended.

Permaculture in a Nutshell. A good start for the basics.

Urban Permaculture. Contains a checklist with useful ideas.

Designing & Maintaining your Edible Landscape Naturally. Contains a checklist with some useful ideas (p91)

'Design bite' , regular feature in Permaculture Magazine, Permanent Publications – Recommended.

'Designing Sustainable Human Environments: A Global Necessity', Harrison, L, Article from Ecopolitics V, Proceedings, Ed Ronnie Harding – Recommend if available.

Permaculture Principles

Joanne Tippett

Objective

To give students a solid understanding of permaculture principles and how they can be applied to achieve long-term sustainability.

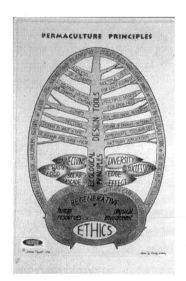

Learning outcomes

By the end of 10 sessions on principles, students should be able to:

· Explain the use of permaculture principles in the process of design.
· Show how these principles can be used at any level of scale and in a multitude of situations.
· Practise applying each principle in a practical way.
· Show the links between principles.
· Apply all of the principles in design exercises.

Context

The first five days are structured around showing links and connections between the principles, and of unfolding ideas in layers. Techniques and topics appropriate to the group's interests, such as forest gardening or water harvesting, are briefly introduced in order to reinforce principles and ground them in practical examples. There is a three hour review of the principles at the end of this week, half of which is spent in small groups mind-mapping the main themes learned up to that point. (See my Example Course Format on page 58.)

Duration

See below

How I teach these sessions

There are 10 sessions focused directly on permaculture principles in the first half of the course, each lasting 90 minutes, with a five minute break in the middle.

The teaching of each principle is roughly broken down into:
20-40 minutes group exercise related to the principle
10-15 minutes group discussion of ideas and issues related to the exercise
5-10 minutes direct teaching of the principle, using visual aids
5-10 minutes of slides and case studies of the principle. This may include observation of the environment outside the classroom
5 minutes discussion and links to other principles.

Below are three examples of exercises I use when teaching the principles.

1 Relative location & connective strategies

Objective

To understand the spatial nature of design and how this influences relationships, use of the land and material and energy flow.

Learning outcomes

· Apply the knowledge gained from Analysis of Elements exercise to the concept of arrangements of elements in space.
· Name different types of connective strategies and how these can be used to enhance productivity and sustainability in design.
· Observe the surrounding area for examples of relative location, either in its absence or application.

Duration

20 minutes

Context

After analysis of elements, before zones.

Resources needed

Coloured pens, scissors, blu-tack, tape, string, blank pieces of card, and pieces of card with various elements written or sketched on them, eg ducks, cow, orchard, flowers, pond.

Activity

Best done outside, or on a large floor-space, with the whole group standing in a circle. Ask the group to place elements in relative location to each other, using the information of inputs and outputs to determine beneficial relationships. I use elements generated in a previous Analysis of Elements exercise. Add any connective strategies to join the elements. Provide strips of paper for this. Don't include any elements such as house, kitchen or person in this exercise, to prevent getting stuck on a discussion which you want to have later when teaching zones. Define the term guild. At this stage, attach identifiable guilds together, using strips of paper and blu-tack, writing any connective strategies on the strips. Keep these guilds for use in the upcoming zones exercise.

2 Zones

Objective

To extend the understanding of the spatial nature of design to understanding patterns of land use and access.

Learning outcomes

· Analyse an area of land in terms of zones of use.
· Discuss patterns of access for various elements.
· Place elements in zones on the site map.
· Talk about types of access and paths on the land and the way this relates to zones.
· Discuss how this knowledge can be used in design.

Duration

30 minutes

Context

After relative location.

Resources needed

Assembled cards of guilds from previous session, blank pieces of card, coloured chalk, coloured pens, scissors, blu-tack, tape, string, area of concrete or flooring on which you can use chalk (if this is not available use large sheets of butcher's paper).

Preparation

With coloured chalk, draw a simple farm layout on the concrete, large enough for the whole group to be able to walk around. I include house, boundary, path, hill, spring, etc.

Stage 1 – Marking the zones

15 minutes

With the group in a circle around the map, hand a long piece of string to two people and ask them to mark what they think would be zone 1, where you go often, easy access. Once people are happy with what has been marked in string, hand another person a coloured piece of chalk and get them to chalk in this zone, writing '1' in the middle. Continue for each zone. Emphasise the fact that zones are not perfect circles, but are a tool for planning where to place elements in relationship to how often you visit them. Zones may be in pockets, eg there may be a pocket of zone 1 at the entry gate.

Stage 2 – Placing elements and guilds

15 minutes

Students should take the guilds, previously assembled in the relative location exercise, and place them in the zones which have been marked on the farm layout map. Encourage discussion about how we use the land and how elements work in relationship to centres of activity. Don't let the discussion digress into too much detail, instead, reiterate the basic principles. Every time someone places an element or guild, encourage discussion among the group. As this is going along, you can hand out some more elements cards. You may sketch appropriate ones in order to lead the discussion. Ask at the end of the session why this is a useful exercise.

3 Sectors (see photo)

Objective

To gain an appreciation of the way in which design can influence and use the energies and material flows through a site.

Learning outcomes

· Show how potentially destructive energies can be deflected.
· Demonstrate how potentially useful energies can be channelled and maximised.

Duration

35 minutes

Context

After teaching about the water cycle, hydrosphere and the role of trees. A good link is to do a windbreak role play (see below), to lead into the sectors exercise. The principle of sectors is later linked with the principle stacking in space and time through the topic of agroforestry, and with the principle cycling of energy and materials through the exercise of designing energy efficiency in a household.

Resources needed

Scarf, string, small rectangular box, torch, twigs, bits of cloth.

Stage 1 – Windbreak role play

10 minutes

(adapted from the Manual for Teaching Permaculture Creatively, Ref 1.)

Go outside, or to an area where you can run around. Say 'I am going to be the wind.' Make appropriate noises and wave a scarf around. 'I want you to form a windbreak to stop me.' Act the wind and see what they do. Then ask how it could be made a better windbreak. Include more people, until everyone is pretending to be a windbreak. In order to direct the process towards forming a classic windbreak you may need to be a bit difficult and insist on being turbulent at straight edges!

Stage 2 – Marking wind sectors

10 minutes

Preferably done outside, so students can feel the sun and wind. With the group, mark the centre point of an imaginary map on the ground. Hand out string. Mark the co-ordinates of north and south. Use a bit of gold cloth for the sun at midday in the summer. Ask someone to mark sectors for the prevailing hot, dry summer winds, then the cold winter winds. (If there is no-one on the course with local knowledge, make sure you know the directions). Give out pieces of cloth or twigs, and ask people to place windbreaks, evergreen for winter winds, deciduous for summer winds. Encourage discussion. If necessary, remind people of the shapes of the windbreaks from role play.

Stage 3 – Marking other sectors

15 minutes

Ask someone to show, in a line beginning at the centre of the map, where the sun rises and sets in both summer, and winter. Use different coloured string to mark the sun sectors. Give someone a rectangular box, which represents a house, and ask them to place it somewhere, considering orientation to maximise sunlight heating in the winter. Demonstrate different sun angles with a torch, then the use of trellises and overhangs for shade in the summer, which won't also block out the winter sun. Have people look at nearby houses and question whether or not they have been placed with good sector planning. Discuss any other sectors, eg fire, pollution, noise, and how they can be dealt with.

Link to ethics

The principles provide a practical way to achieve permaculture ethics. As I teach the principles, I refer back to the ethics to encourage more discussion. I link this teaching to discussion of how permaculture principles can move towards achieving the system conditions of sustainability elucidated in The Natural Step model (see pages 89-90).

Further activities

I use two design exercises on the course to consolidate the teaching of the permaculture principles. One is a short, unstructured exercise at the end of the first week, which provides an opportunity for people to see how well they have understood the permaculture principles. The second week of the course focuses on a more structured and wide-ranging design exercise.

Bibliography

1 The Manual for Teaching Permaculture Creatively

The Natural Step

Joanne Tippett

The Natural Step (TNS) is a consensus-based approach to sustainability. It was started in Sweden in 1989 by Dr. Karl-Henrik Robert, a leading cancer specialist. Based on scientific principles, it contains a clear description of the conditions necessary to achieve sustainability. As a tool it reduces confusion, cutting through seemingly conflicting information in the sustainability debate. It offers a framework for strategic planning, and can be used as a compass for navigating step-by-step towards long-term ecological sustainability.

TNS is a useful tool for optimising permaculture education and design. It provides a foundation for the practical implications of the ethical basis of permaculture. Permaculture can then build on this foundation with descriptive tools and principles to determine how sustainable design will function and look in different situations.

It has been used in Sweden as a tool for education and strategic planning with over 60 corporations, 50 municipalities and 17 professional networks. A public awareness programme was targeted at the population of Sweden, with an educational pack sent to 4.3 million households and a widely publicised television programme on 30 April 1989. TNS organisations are being established in seven countries, as the beginning of TNS International.

The scientific principles of The Natural Step are:
· Matter and energy can neither be created nor destroyed (the law of conservation).
· Matter and energy tend to disperse spontaneously (second law of thermodynamics).
· Biological and economic value (quality) is in the concentration, structure and information content of matter. This quality is the only thing we consume.
· Sun driven processes are essentially the only net producers of material quality on earth (photosynthesis).

The systems conditions for sustainability:
(Copyright of The Natural Step, Ref 1.)
When the society is sustainable, the functions and biodiversity of the ecosphere are not systematically undermined by:

    I increasing concentrations of substances extracted from the lithosphere.

    II increasing concentrations of substances produced in the technosphere.

    III physical impoverishment from manipulation or overharvesting.

Together these system conditions give a framework for ecological sustainability. It implies a set of restrictions within which the sustainable societal activities must be incorporated. Based on that reasoning, a first order principle for the society's internal turnover of resources can be formulated:

    IV when the society is sustainable, the uses of resources is fair and efficient enough to meet vital human needs everywhere.

Bibliography
1 The Systems Conditions for Sustainability – a Tool for Strategic Planning, Holmberg, J & Roberts, K H, The Natural Step, 1998.

Further research

In the UK, Forum for the Future is the licensed operator for The Natural Step. They can be contacted at:

The Natural Step UK, Thornbury House, 18 High Street, Cheltenham, Glos, GL50 1DZ.

Tel: (01242) 262744. Fax: (01242) 262757.

Email david.cook@tnsuk.demon.co.uk

Websites

Many useful webistes exist. Go to www.permaculture.org.uk for the latest links.

⇨ Principles of Permaculture – Cycling

Mike Feingold

○ Objectives

To understand natural cycles, to look at wastes and their uses, and to examine our attitudes towards waste and how to turn problems into assets.

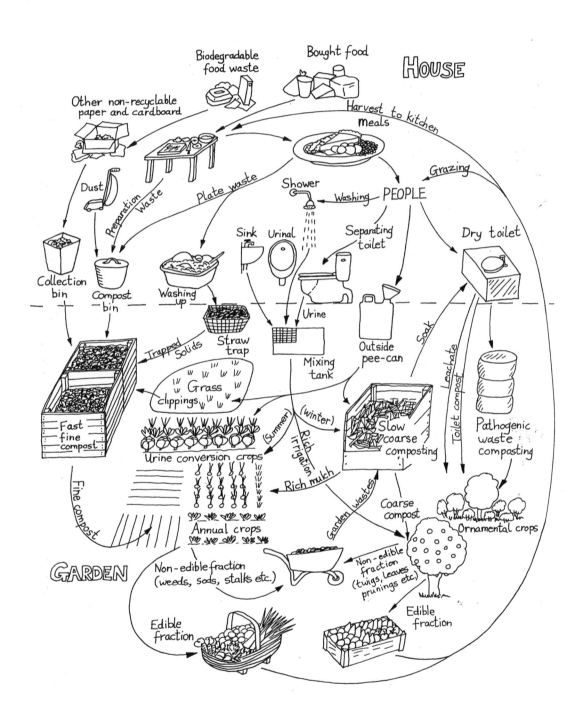

Learning outcomes

By the end of this session students will be able to:

· Explain the importance of cycling.
· List cycling strategies.
· Take action to reduce waste and increase cycling.

Context

Day 2, during which I go through the conscious application of principles. It follows traditional systems, which gives examples of cycling, eg animal wastes to the land, crop residues to animals etc. I will also have done a slide show on how communities live around the world.

Duration

At least 90 minutes

How I teach this session

Interactive talk

30 minutes

I start with the cycles all around us. Nature moves in cycles and I ask the group to think of some of them, eg lunar, seasons, day/night, menstrual, water, nitrogen, seed-plant-seed etc.

I then look at some cycles in more detail, showing how we can apply the principle practically. Example, kitchen waste:

· Bin to landfill = disaster, waste and pollution.
· To compost = pat on the back.
· To animal to manure and yields = more cycles, good.
· Manure of animal to earthworm to chicken to biogas to land = even better!

I also give examples of cycling within local economies.

It is crucial that people understand that each time nutrients go through a biological cycle, there is an increase in yields. Increasing the number of cycles, and their speed, therefore increases fertility, yields and wealth.

Activity – observational walk

15 minutes

· Make a list of all the items of 'waste' that you see. Which of these wastes could you turn into an asset? What materials do you recycle?

After the walk, students are asked to contribute some points to a whole group discussion.

Who recycles what?

10 minutes

There are normally some very keen recyclers, so this gives them an opportunity to tell others what they do, and how.

Design exercise in small groups

15 minutes

What are the important aspects of a recycling facility?

Case study – slide show

30 minutes

I often end with a slide show showing the Centre of Alternative Technology as a case study, including reed beds, compost toilets, composting etc.

Most of the following points will be covered during the session:

·   Outputs of the system are used as inputs, or become pollution.

·   Fertility is a function of cycling. Cycling builds up natural wealth.

·   Cycling is related to diversity and making useful connections.

·   Trees and plants are essential cycling systems.

·   Potential yields increase as the rate of cycling increases.

·   Cycling water minimises the use of resources. Water has many 'duties' to perform,
    it is multi-functional.

·   Nutrients can be cycled through compost systems, chickens, pigs, insects and fish ponds.

·   Personal responsibility is essential for minimising waste, and increasing cycling.

·   7Rs – Rethink, Redesign, Reduce consumption, Reuse, Repair, Refuse (be selective, return
    packaging), Recycle.

·   Find second use, or turn back into original, eg second use for silt is bricks, and a second use
    for the hole is a fish pond.

·   Design materials that can be cycled – 'Precycling'.

·   Biodegradable, what this means and what it doesn't.

·   Restrict materials that don't break down or cycle beneficially, eg DDT, lead batteries,
    radioactive materials.

Further activities

·   During the course, make sure all 'wastes' are recycled, composted and minimised.
    Involve students in this.

·   Carry out a practical composting activity.

·   Consider the inputs and outputs of cities. How can these 'import monsters' be redesigned
    to minimise adverse environmental impact?

Principles in Pictures

Maryjane Preece

○ Objective

To help students remember and understand the principles, through the use of simple drawings.

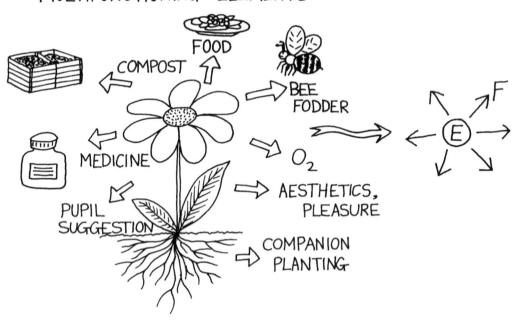

○ Learning outcomes

By the end of the session students will be able to:

· Draw simple pictures which encapsulate the meanings of the principles.
· Review the principles quickly and easily.

○ Context

During and after principles sessions.

○ Duration

Initial explanation: 5 minutes
Homework or group work: 45 minutes
Feedback: 20 minutes

○ How I teach this session

Initial explanation

5 minutes

"By making drawings of the principles, we can remember them more easily. A good picture will remind us of all the important points about each principle and the process of making them in the first place will help us to think about them in a new way and understand them better."

During principles sessions

I present my own pictures of the principles as they come up. This gives useful visual input to complement talking and exercises. Students can copy these as they make their notes.

Homework

30-45 minutes

Since drawings made by the students themselves are more likely to be remembered, I ask them to prepare their own drawings for the next day as a piece of homework. This can be done by asking each student to do one or two, and covering the range of principles between the group, or by asking all the students to do all of them.

Next day review

20 minutes

I make time on the following day for their drawings to be presented and discussed. They can be shared between the group and individuals can jot down the ones they like most as their own personal memory joggers. This also acts as a review for the principles session.

This technique can be applied to many other topics and sessions.

Principles Review Exercise

Andrew Goldring

Objective

To review and memorise the principles using song, dance/movement, drawing and story.

Learning outcomes

By the end of the session students will be able to:
· State and remember the principles easily.

Context

One day after the principles session(s).

Duration

45 minutes

How I teach this session

Resources needed
Paper, crayons, pencils and pens, different rooms or spaces for groups to work separately, enough floor space for presentations.

Explanation and grouping
After a quick explanation of what we will be doing, I let people get into groups depending on their preference, ie song, dance/movement, drawing, story.

Group activities
25 minutes
Now ask the groups to incorporate the principles learnt the previous day into the activity they have chosen. Make sure that each group has their own space to do this in.
During this time I look to see how each group is doing, but don't interrupt or get involved unless asked.

Presentations
15 minutes
Ask each group to present what they have done.

Closing round
5 minutes
Invite students to share something they have learnt or observed about the principles during the session.

Further research

Use your memory, Buzan, T, BBC Books, 1994. Chapter 11.
The Manual for Teaching Permaculture Creatively.

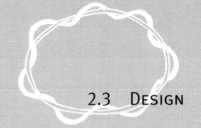
Broad objectives for design (over a number of sessions):

· To present a range of methods and processes that can be used when designing.
· To allow opportunity for students to practice using some of these methods and processes.
· To present a range of processes that include people (clients and communities) in the design process.
· To develop simple skills in mapping, surveying and drawing.
· To bring together all these skills within a final design exercise.

 # Design Methods and Processes

Patrick Whitefield

○ Why I include this session

Some comments from students on the value of using formal design methods:

· Design methods act as a check on less structured work.
· Zone, sector and elevation analysis catches things which would otherwise be missed.
· McHarg makes you face up to constraints.
· The obvious things are not always the best.

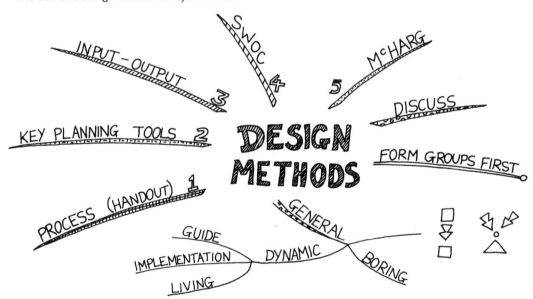

○ Objective

To give a sufficient introduction to design methods for the design exercise.

○ Learning outcomes

By the end of the session students should be able to:
· Do the design exercise.
· Give an account of the various design methods which can be used.
· Have a stab at choosing suitable methods for a particular design task – they will be much better able to do this after the Main Design Exercise.

○ Context

Either: a) before the Base Map Exercise, or b) before the Main Design Exercise.
(a) has the advantage that the students get an idea of design methods before they start the sequence of exercises which make up the Design Process. (b) has the advantage that the design methods will be fresh in the students' minds when they do the Main Design Exercise.

If there has been a Short Design Exercise earlier in the course I invariably choose (b), as they will have had a brief introduction to the design process already. Otherwise I choose whichever seems most convenient in the individual circumstances of the course.

○ Duration

45-60 minutes

○ How I teach this session

Before the session starts the students have formed themselves into their design teams. At the start of the talk I tell them that at the end they will be selecting two of the methods I am about to present to use in this exercise.

Part 1: Talk

20-35 minutes – the longer time if there has not been a Short Design Exercise earlier in the course.
General:
· The concept of sustainable design (see 'Principles' session, pages 73-77.)
· Beware wild designing; the best designs often look boring.
· Permaculture design is dynamic, because:
  · living things grow and change through time;
  · the design will be implemented in stages;
  · what's usually wanted is guidance through a process rather than presenting
    a complete picture of 'how it's going to be'.
Outline five design methods:
· The design sequence (BREDIME):
  · covered already if we had a Short Design Exercise
  · not a separate method, but the context for all others
  · handout outlining the design sequence.
· Key planning tools – zone, sector, network, elevation:
  · already explained in the 'Principles' session
  · must always be part of your consciousness during permaculture design,
    but can also be used as a specific design tool.
· Input-output analysis:
  · as illustrated in the example of the chicken
  · again, both a constant awareness and a specific tool.

- SWOC, Strengths, Weaknesses, Opportunities and Constraints:
  - can be done for the site as a whole or for specific parts or elements.
- The McHarg Exclusion Method (Ref 1):
  - illustrate with McHarg's original example
  - discuss what kind of design jobs it may be useful for.

The McHarg Exclusion Method

Ian McHarg is a Scotsman who has spent most of his working life in North America as a professor of landscape design. He was once asked by a group of local residents to support them in objecting to the route of a proposed road. In working towards a proposal for an alternative route, he came up with his exclusion method.

The basis of his method is to ask not where something should go, but where shouldn't it go. A base map is drawn and a series of transparent overlays are prepared, each one mapping areas which are excluded for a specific reason. In his original work on the road proposal the subjects for overlays included: too near to residential areas, forest, areas of wildlife value, marsh, and areas incurring extra expense, eg a bridge.

When all the overlays are placed over the base map at once any area which remains blank is ideal, and areas which have the least constraints can be considered if the blank area is not sufficient.

The method can be used for placing new structures or plantings in the landscape, including: settlements, individual houses, farm buildings, new woodland and orchards.

See also 'Design by Identification of Limiting Factors' on pages 113-114

Methods I don't include because I haven't found them useful in practice:
- Random assembly.
- Flowcharts.

Part 2: Discussion
15-20 minutes

Ask the students to get into their design teams and select two methods which they will specifically use on this job. This doesn't mean they can't use more than two, but they are asked to assess which two seem most suitable in this particular case. They will be asked to report back on them afterwards.

○ Bibliography

1 Designing with Nature

○ Further research

Permaculture: A Designers' Manual, Chapter 3
How to Make a Forest Garden

Design Methods and Processes

Simon Pratt

○ Objective

To indicate the wide range of methods available, with one or more to be experienced as part of the design exercise.

○ Learning outcomes

By the end of this session students will be able to:
· Use one or more design methods.
· Understand that there is more than one way to design.

○ Context

The middle of the course, before the design exercise.

○ Duration

45 minutes

○ How I teach this session

I present some of the following as a toolkit to dip into. I wouldn't teach them all in one session, or even in one course. Some of the processes may have been presented earlier, if an appropriate moment arose, in which case I would quickly refer to them again in this session. I would mainly present a process on the board with some explanation, giving time for questions. I make sure that students do at least one practical activity during the session.

The design processes are discussed again before the main design exercise, which is the students main opportunity to put them into practice. During the evaluation of the designs, I ask how they found using the different design processes.

· Design sequence I understand as 'BREDIME' or 'OBREDIME':
  (**O**bservation – included in all stages)
  Boundaries
  Resources
  Evaluation
  Design
  Implementation
  Maintenance/Monitoring
  Evaluation

· Another description of the design process I find helpful:
  Information phase
  Analysis phase
  Design phase
  Management phase – splits into implementation and maintenance

- Input-output analysis. I sometimes use the Parable of the Chicken (also known as 'A Tale of Two Chickens', see pages 183-186) as a warm-up exercise early in the course, so a reference back to this session would be useful at this stage. Input-output is a process which can help to measure the viability of a plan. I sometimes ask students to include in their designs answers to the following questions:
  1 What are the costs involved in implementing the design? £, time, available resources etc
  2 Ditto for maintenance.
  3 What yield will the system produce?
  4 Where is there a shortfall in resources as things are at present?

- SWOT – Strengths, Weaknesses, Opportunities, Threats.

- Random assemblies. This can help students to make connections between elements in a system and lead to creative thinking, where what seem to be silly connections lead to a deeper understanding. Also a good participatory exercise, either for the whole group or in small groups, ie it precedes the 'hard plan'.

- Use of pattern language. Pick one of the patterns and ask students to discuss. Alternatively test a design or an existing situation against the pattern language. Tell students it's OK for them to write their own patterns and thus help build up the language (Ref 1).

- Yeoman's Relative Permanence scale (Ref 2)
  1 Climate
  2 Land shape
  3 Water
  4 Roads
  5 Trees
  6 Buildings
  7 Fences and boundaries
  8 Soil

- ASME process of comparing different ways of doing things (from industry via Andy Langford)
  - **O** operation
  - **⋯⇢** transport
  - **D** delay
  - **S** storage
  - **I** inspection

  For example, production of fence posts
  a from softwood plantation using conventional process
  - **O** cut/cross
  - **S** pile
  - **O** load
  - **⋯⇢** sawmill
  - **S** pile
  - **O** peel
  - **O** point
  - **I** check
  - **S** pile
  - **O** load
  - **D** process/dry

S pile

O load

┅⟩ delivery

O use

b from sweet chestnut on site

   O cut/cross

   O point

   O char ends

   S pile

   O load

   ┅⟩ use

· Ecological footprint as a design tool. It can be used to check the ecological impact of different designs and existing situations. Our Ecological Footprint (Ref 3) is filled with practical examples, and can be used to generate exercises during the course. Other related methodologies can also be mentioned, eg environmental impact assessment.

· Kourik's methods, written before the Designers' Manual, provide an interesting parallel to Mollison's ideas, there being many similarities. Good for garden and landscape design, it contains Golden Rules of Edible Landscaping (Ref 4, p91), and a humorous collection of Kourik's practical experience, which are similar to Mollisonian principles. He refers to Roger von Oech's 10 causes of mental blocks, which can help with creative thinking and designing (Ref 5, p84).

· Graham Bell's checklist can be used to test if a design is on the right track (Ref 6, p210).

○ Bibliography

1 A Pattern Language

2 Water For Every Farm

3 Our Ecological Footprint

4 Designing & Maintaining your Edible Landscape Naturally

5 A Whack on the Side of the Head, von Oech, R, Warner, 1990

6 The Permaculture Way

○ Further research

The Basics of Permaculture Design

The Selection of Design, Clegg, G, CUP, 1972

Design for the Real World, Papanek, V, Thames & Hudson, 1983

The Green Imperative, Papanek, V, Thames & Hudson, 1995

⇨ Mapping

Simon Pratt

○ Why I include this session

Designing a piece of land is almost impossible without some basic mapping skills.

○ Objective

To explain possible sources of base maps and experience the practical exercise of producing a map.

○ Learning outcomes

By the end of this session students will be able to:

· Explain how to source base maps.
· Produce a simple map.
· Explain that the map is not the territory.

○ Context

There's a fair amount of overlap with the surveying techniques session, so it's good to combine them or run them close together in the course.

○ Duration

Source of base maps: 45 minutes
Base map exercise: 180 minutes
Self-guiding tour model: 45 minutes

1 Source of base maps

45 minutes

How I teach this session

I find it is good to have several examples of different maps available for students to handle. Ideally a map of the course venue and/or design site should also be available.

Brainstorm

Start the session with a brainstorm on maps. Try to draw out the whole range from plasticine models to photos, video, notebooks and Geographical Information Systems on computers. There are billions of ways of mapping, and this exercise really opens up the territory.

Main points

· Clearly state that the map is not the territory.
· Always use maps in association with a site survey.
· Maps rarely show financial, social, cultural or ethical details.
· Remind students that it's OK to start with a base map that someone else has done the survey work for!
· Give sources of base maps and information:

- Ordnance Survey 1:2500, 1:10 000, 1:25 000 and 1:50 000 scale maps are available for both urban and rural areas. A 1:1,250 is available for urban areas only. It's useful to have examples of different scales available. Briefly discuss the advantages of each scale.
- A computerised map service, available from local OS agents, can provide a print of any area you want, at any scale between 1:200 and 1:5000. This is very useful where your land is on the boundary between two or more OS sheets.
- Drift (Surface) and Solid (Underlying) Geology Maps.
- Soils survey – Ministry of Agriculture Soil Maps.
- Aerial photographs – OS, RAF, Fairey Surveys and other commercial sources.
- Local library and record office for tithe maps and historical maps showing valuable information on previous uses.
- Ministry of Agriculture Land Classification Maps. These are graded according to usage; seven grades plus 'ungraded'.
- Land Utilisation Maps – historical agricultural uses.
- Forestry Commission Woodland Maps.
- Admiralty Survey maps.

Exercise 1

Discuss contours and if possible do this in a sandpit.

Exercise 2

Ask students to draw a landscape profile from simple contours.

2 Base map exercise

180 minutes

How I teach this session

Opening

If not covered previously explain the various approaches and use of overlays.

Different ways of organising groups

- All prepare the same map, but swap results around either half way through or at end of session, learning from different approaches.
- Groups prepare maps with different types of information, eg water, vegetation, buildings and structures, boundaries, access, soil, all combined at end as overlays.
- Two different groups, eg men and women. This always throws up really interesting points of view.

If possible use the results of this exercise in the main design exercise.

Most groups are able to complete a fairly extensive survey of the design site with little or no preparation, but the teacher must be available to assist.

3 Self-guiding tour model

45 minutes

How I teach this session

I used this exercise on Skye's Advanced Teacher Facilitation Course at Redfield, September 1997. Three groups of about five people were given a different route to follow on the same base map and asked to collect samples of what they found along the way. They were also given a simple question to answer, eg 'Can you find some recent coppice re-growth?'. Meanwhile I prepared a large chalk drawing of the site and they were asked to place found objects on the plan. This could be done inside on a large piece of paper. This is a quick way to build up a picture of the site while developing map-reading and

interpretation skills as well as plant identification.

This can be used on the venue site or for off-site visits. No more than 30 minutes are needed to follow the route, and 15-20 minutes for map-making and informal discussion of what they found.

It can be used to help prepare the base map and/or for an observation exercise, familiarisation with the site, or site visits.

○ Further research

The Permaculture Way, Chapter 11.

British Regional Geology, British Geological Survey, The Stationary Office. Twenty regional books.

Phone the MAFF helpline on (0645) 335577 for information on Agricultural Land Usage/Land Classification maps.

⇨ Surveying Techniques

Simon Pratt

○ Why I include this session

Building and using simple surveying tools helps to demystify the process for students.

○ Objective

To experience practical surveying techniques and construct simple surveying equipment.

○ Learning outcomes

By the end of this session students will be able to:
· Construct their own surveying tools.
· Survey an area of land.

○ Context

There's a fair amount of overlap with the mapping session so it's good to combine them or run them close together in the course.

○ Duration

40 minutes

○ How I teach this session

This session works best as a very practical exercise, constructing the equipment from scratch and then using them in the landscape. Depending on the group, they can either have an introductory talk lasting five minutes, or simply be given the materials and asked to get on with it! (Make sure you have a suitable site available.)

I point out that surveying is more important for potential house sites than broadscale landscapes. And that in permaculture design, observation is more important than accurate surveying. I answer questions as they come up during the surveying exercise.

The two simplest tools to construct are the builder's level, also known as a water gauge or bunyip, and the A-frame.

Good preparation is essential for this session.

a) Builder's level

Two sticks approx 50 x 25 mm and about 2 m long.
About 20 m of clear plastic hose 10-12 mm diameter.
Two corks or stoppers to fit in the end of the hose.
Four clamps or tape to connect the hose to the stakes.
Water finds its own level, so when the hose contains water the two ends will always show the same level. The difference between measured marks on the two stakes gives the level between the two points. Very useful in laying out swales (level) or drains (1 in 500 fall or greater).

b) A-frame

Three pieces of wood (bamboo is ideal) are fixed together to make the basic frame. A plumb bob hanging from the centre shows the centre point. Used by swivelling alternate legs until the plumb bob settles at the same position, this method has been used for centuries for laying out contours or drains.

It is worth differentiating between the benefits of these low-tech tools. a) can level or set slope over greater distance at one go. b) is easier to make and can be used by a single person.

Further activities

If time permits and students express an interest, a brief mention of other techniques can include:

· Distance measurement – pace counting, tape measure (50 m), surveyor's wheel and optical range finder.
· Levelling – builder's level, plane table, A-frame, dumpy level and clinometer (if you can do basic trigonometry) – or setting angle of slope.
· Locating points – use of compass, sighting from fixed points, sighting compass.

Above are surveying techniques to produce a conventional base map of physical features. As permaculture designers we need additional information about the land such as soil and drainage, so appropriate tools like pH meters, soil augers and salinity meters should get a mention. Infiltration rates can be tested by digging a hole and seeing how long it takes for water to drain away.

Further research

Permaculture: A Designers' Manual, p232

The Basics of Permaculture Design, Chapter 5

Mapmaking for Orienteers, Harvey, R, British Orienteering Federation, 1990

Input-Output Analysis

Andrew Goldring

○ Objective

To explain the Analysis of Elements design method, and, using a simple group exercise, develop this into an opportunity for students to uncover some permaculture principles.

○ Learning outcomes

By the end of the session, students will be able to:

· Use the Analysis of Elements design method.
· Explain why observation is crucial to the design process.
· Explain what concepts such as pollution, extra work and relative location mean in a permaculture context.
· Explain why creating beneficial relationships is crucial to permaculture design.

○ Context

Early on in the course, usually the first day.

○ Duration

60 minutes

○ How I teach this session

Introduction

I introduce the session by drawing up a large person with accompanying text (see left). I use a person rather than a chicken, because people tend to know more about people than chickens, and the mapping of needs/inputs, intrinsic characteristics and products/outputs comes in useful during the Zone oo session.

Whole group exercise

Asking students to shout out suggestions, I map inputs/needs, move to outputs/products and finish with intrinsic characteristics. We repeat this, until a rich picture has formed.

Question

I ask students how this method of analysis could be useful in design, and map up all the suggestions, without comment.

Small group exercise

Now I get out four pre-drawn grids (see next page). I put up a very large version at the front, which I use to give the instructions for the exercise and later to collate the results. Then I split the students into three groups and give each group a grid to work on.

I ask each group to spend five minutes filling in the outputs for the different elements on the grid, just as we did for the human. Next I ask them to spend five minutes filling in the inputs. I tell people that the lists do not need to be extensive, so they should concentrate on getting two or three for every element first, and spend any remaining time adding to these.

Now I ask them to spend five minutes matching up inputs and outputs and making suitable connections.

When they have finished I spend about 10 minutes collating all the inputs and outputs that the groups came up with. This can get tedious if the group is large, so it could also be done by students coming up to the main grid and filling them in.

Now I ask each student in turn for a connection, which I map up. This continues until everyone has made a contribution, and most of their connections have been filled in, or until the grid becomes completely unmanageable and looks like a plate of spaghetti!

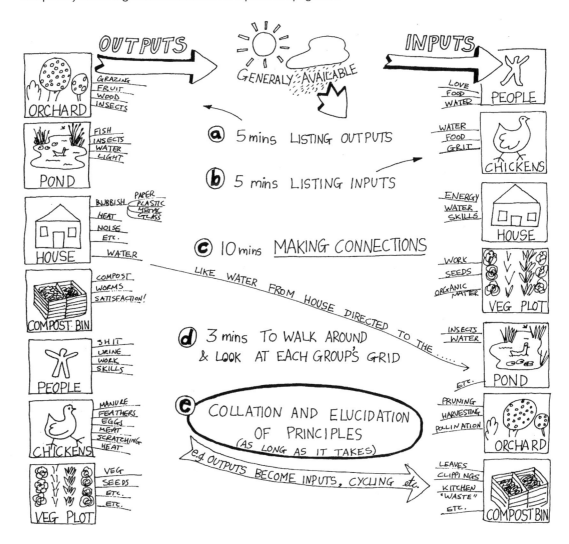

Observations

At this stage I ask students for their observations and what they have learnt and map them all up. Concepts that this exercise should help to point out include:

· pollution

· extra work

· multiple functions for each element

· importance of relationships

· relative location

· cycling

· automatic systems

By referring back to the grid, and asking good questions, each of these concepts can be elicited from the group. Often all that is needed is the rephrasing of contributions so that they fit 'textbook' definitions.

OHP (if available)

At the end of the session I present the diagrams from the Designers' Manual (Ref 1, p24 & 25) which shows a comparison between the industrial process of creating eggs, and the permaculture method.

Further activities

If there is time, I introduce the concept of Industrial Ecology (Ref 2, p61-67).

Bibliography

1 Permaculture: A Designers' Manual, p24-25
2 The Ecology of Commerce

A Farmyard Somewhere –
An Exercise in Placement and Flow

Simon Pratt [first produced by Andy Langford with notes by Patsy Garrard]

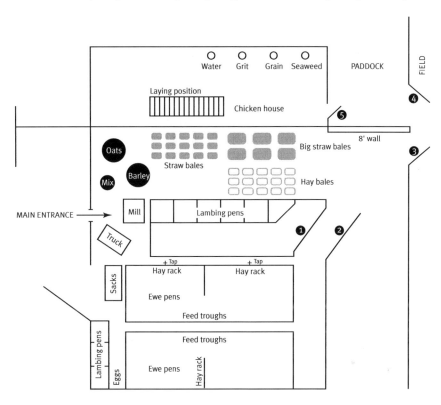

Why I include this session

Flowcharts are particularly useful in designing work places, where people, energy or materials flow from one place to another. They can be critical in reducing time spent on maintenance of existing systems so that new and creative work can proceed. An aim of permaculture design is to change the usual ratio of 80:20 for maintenance:creativity and costs:profit towards a 20:80 ratio.

Objective

To provide a real-world example that can be used to develop design strategies that improve the efficiency of a system.

Learning outcomes

By the end of the session, students will be able to:
· Explain how small changes in the placement of elements within a system can make large differences to its operating efficiency.
· List strategies to improve placement and flow.
· Design a little more confidently!

Context

Best included with other design processes, although it works well as a stand alone exercise.

○ **Duration**

45 minutes

○ **How I teach this session**

Pin a plan of the farmyard to a piece of cardboard or pinboard.

Part 1

10 minutes

Following the list of journeys below, place a pin at each point of the journey described. Put two pins at the bottom of the board, 50 cm apart, and use them to run out 8 m of cotton. Tie the cotton to the pin at the starting point and run the cotton around each point in the journey. When you have finished, measure what's left.

Part 2

15 minutes

Now ask students to relocate things and make other suggestions to reduce journey time. They are not allowed to demolish the eight foot wall!

Part 3

10 minutes

Run out another 8 m of cotton and measure the re-routed journeys. Measure what's left.

Part 4: Evaluation

10 minutes

Spot the difference! ie time saved.

The list of journeys:

Start from the main entrance.

· Check sheep. It is lambing season and you have just arrived at 8 in the morning to take over from the night watch.
· Walk down one alley and up the other, casting a watchful eye to both sides. Nothing doing so on to feed the chickens.
· Feed the chickens. Collect two recycled bags from sack store, take to grain bin area and fill bags from mixture bin.
· Take bags, one at a time, on shoulders (alleys too narrow for trucks) to chicken house via gate 2 (untie baler cord fastener), gate 3 (ditto fastener), gate 4 (ditto fastener) and up to door 5. Close all gates each time you pass through as there are sheep with lambs in various stages of readiness to go out in all areas of the yards, and mix ups (wrong lambs to wrong ewes) are a pain to sort out.
· Open door 5 and quietly pass both bags through, prop them just inside the door and close up before chickens escape. Now fill both feed bins from the sacks and check state of seaweed and grit hoppers. Carry empty sacks back to sack store, collect bag of seaweed from same place and take to chicken house to refill hopper. Observe gate drill! Return to sack store and find several empty egg buckets and return to chicken house, delicately collect eggs from laying stalls and take these to eggs store by the top of the lambing pens. Do be careful with the gates through this operation.
· Now find empty water carriers, say in long suite of lambing pens. Now fill two at taps in alleyway and carry, one at a time, up to the chicken shed. Fill troughs and return carriers to the tap area. Gates!
· And rest!

Do this twice a day before attending to the sheep and then the cattle.

Design by Identification of Limiting Factors

Simon Pratt

Why I include this session

This is a simple but technically rigorous design process, which students can learn easily . It is a quick classroom exercise – good for rainy days.

Objective

To explain the concept of limiting factors in permaculture design and how to use them in the design process.

Learning outcomes

By the end of this session students will be able to:
· Identify limiting factors.
· Explain that limits are more than local topography.
· Use limiting factors as a design process (McHarg Exclusion Method, Ref 1).

Context

This works well as a stand alone session, but can also be combined with the main mapping and surveying sessions.

Duration

Approach 1: 20 minutes
Approach 2: 40 minutes

How I teach this session

Approach 1
Short presentation
15 minutes
Using an OHP, I start with a blank map and add individual overlays to explain the McHarg Exclusion Method. Factors such as steepness of land, solar aspect, risk of flooding, cold winds and frost pockets are mapped as separate overlays and used to determine ideal sites for dwellings, crops, water storage, animals, industry and so on. (Also see box on page XXX.)
Discussion
5 minutes
'What other limiting factors can we identify?'

Approach 2
Short presentation
5 minutes
Explanation as above, but without using an OHP.
Exercise
20 minutes

Set the element(s) to be placed on the landscape. Students start from a blank map and produce their own overlays.

Presentations of work

10 minutes

Each group present their solutions.

The teacher may need to remind the group that limiting factors are not limited to topography and physical characteristics of the site, but can include financial, social, cultural or ethical considerations.

Bibliography

1 Designing with Nature

Zonal Analysis of People and Information

Mark Fisher

Why I include this session

The emphasis in conventional zoning is land use. It is often forgotten that in permaculture the origins and use of information as a resource can also be described through zonal analysis. In addition, my experience of working locally, regionally and nationally on Local Agenda 21, organic growing and permaculture leads me to believe that people and their organisations can also be described this way. I have found that teaching the zonal analysis of information and people has given me a better understanding of how we may individually and collectively design for permanent beneficial change in our society.

Objective

To explore how zonal analysis can be used to describe organisations and information.

Learning outcomes

By the end of the session, students will be able to:
· Apply zonal analysis to information and to people and the way they organise themselves.
· Understand where information comes from and how its value is in its exchange and use.
· Identify the relationship and importance of local, regional and national activity, and how this shapes society.

Context

This approach can be included in the initial teaching of zoning or as a separate session that is returned to later in the course. It links well with Local Agenda 21 and Local Food Links.

Duration

45 minutes

How I teach this session

My approach is to lead the group through stages that build on their understanding of zoning. Question and answer is the key with the occasional observation thrown in as the spur for the next stage.

Interactive review
5 minutes
I start with a review of land-based activities in the different zones before relating the zoning of information to those activities. The simplest analogy to give the group is the distinction between wilderness and Zone 1.

In wilderness, or Zone 5, we are students at the feet of Mother Nature. We observe natural systems and learn lessons that we may use in other zones. In contrast, in Zone 1 we are arranging the natural world to meet our own needs.

Group discussion
10 minutes

I then ask the group how information arises in other zones. It is recognised that it is likely to be through a combination of the interaction between people and other species.

For the zoning of people it is important to readjust the group's perception of distance between zones. In land-based zoning, we are talking in terms of walking distance from our home, whereas in people-based zoning the distances are much further and the zones can be classed as local, regional and national.

Using the group's experience – pairs or small groups

5 minutes

To get the group thinking in terms of people-based zoning, they should be asked for examples of the activities they get involved in locally. This could include connections with voluntary groups that have similar aspirations and outlook, ie community food growing, a Local Agenda 21 network, a local permaculture or organic gardening group, LETSystems.

Parallels can be drawn with a land-based Zone 1 in terms of the frequency and energy input associated with this local activity. It will need to be discussed whether employment is a Zone 1 people activity in much the same way that allotments need this consideration.

Moving through the zones

15 minutes

Having taken on a people-based view of zoning, the session continues by considering how this could be extended to other zones ending in wilderness. Examples should be asked for of people-based activities that members of the group are involved in regionally and nationally. It is also important to see what effect regional and national activities have on our local activity. This could be national and regional government and how it affects local government in the same way that national action on Agenda 21 affects Regional and Local Agenda 21. Another approach is to look at food supply and distribution and contrast the merits of a local food economy with the food miles associated with regional, national and international distribution systems.

Making connections between zones

10 minutes

The ability of non-governmental organisations to respond to local needs is an issue rising in importance as more people engage with Local Agenda 21 processes. The analogy of an information (wildlife) corridor stretching from national to local zones, is one way to show how national organisations can be supportive of local action. The group should explore this with examples suggested by the group or identified from Community Works! (Ref 1).

○ Links to ethics

Zoning of information and people's activities aids our understanding of how we individually and collectively organise and integrate People Care with the other ethics.

○ Bibliography

1 Community Works!, New Economics Foundation, 1997. A listing of community activity and some of the regional and national organisations that support it.

○ Further research

Permaculture: A Designers' Manual, Zoning of Information and Ethics, in Chapter 3, and schematic for an individual's probable relationship in space, Chapter 14, p516

PMI Thinking Key

Andy Goldring

Why I include this session

Technical solutions exist for most of the problems facing us today. I believe that the major limiting factor to progress towards sustainability is political, cultural and personal will to actually make the changes required. If we are to tackle this problem we must be able to deal with people's perceptions in a way which respects the input and integrity of each participant and ensures maximum thinking contributions by all.

Objective

1 To show that there is a problem with our perceptions that needs to be addressed.
2 To present a solution, and give students the opportunity to learn how to use it.

Learning outcomes

By the end of the session students will be able to:
· Use a simple model which shows how the brain works.
· Explain why our perceptions are often narrow, and not actually based on thinking.
· Use the PMI thinking tool.

Context

Usually right at the very start of the course. This has a number of benefits:
· People can see from the start that their input is valued, valuable and actively listened to.
· Everyone has an opportunity to speak.
· People realise that we can hold different views, contribute them without any need for argument, and develop a richer picture between us than we can have on our own.
· It is simple and everyone has a go at it during the rest of the course, thus developing practical consultation skills.
· It can be used at the end of each day. With the information it produces, I can adjust my behaviour, pace, schedule, classroom layout and so on very quickly. Thus students can help in the ongoing design of their learning environment. This also helps me to quickly develop an understanding of what types of activities the group finds most valuable.

Duration

120 minutes

How I teach this session

Logic Bubble (the problem)
50 minutes
This is a 'brain experiment' devised by Edward De Bono (Ref 1) which shows how the brain operates as a self-organising information system. Books on physiology and anatomy refer to the logic bubble as a 'reverberating circuit'.

Introduction – talk

5 minutes

· Our perceptions shape our world. An understanding of how the brain works will enhance our ability to bring about positive changes.

· We will do a simple (not simplistic!) experiment to see how the brain works, and then test it against our own experience.

· After the break we will look at a tool that can be used to enhance our thinking processes.

The experiment

10 minutes

I follow my own instructions on a flipchart as I am giving them out, so that students can see as well as hear what to do. I check throughout that students are doing it right by walking around and looking at their paper, helping where necessary.

· Each student gets a piece of A4 paper, landscape format, and divides it down the middle with a line. Label one side A and one B.

· Place some dots randomly on side A, then on side B. Draw a small circle around each dot for clarity. These are the neurons or 'brain cells'.

· Now link them using lines. Each small circle must have at least two lines connecting it to others. These represent neural pathways.

· Now label two of the lines as they come out of each circle with a 1 and a 2. Do not label the third or fourth lines. Make clear that the line could have a number 1 at one end and a number 2 at the other.

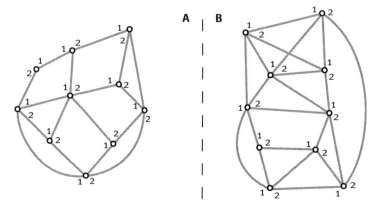

· 'Now we are ready to start the experiment', I tell them. The rules are as follows: 'Always leave a circle by a 1 unless you came in by a 1, in which case leave by a 2 (ie you can't go back on yourself). Choose any circle on side A, place a pen on it and mark a path from circle to circle.'

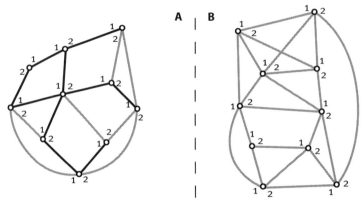

· I leave people to it for a few minutes and then ask if anyone has noticed anything. I suggest they try starting from another dot. Again leave them for a minute or so and ask again if anyone has noticed anything. Now ask them to try side B. After a couple of minutes I stop everyone again and collate their observations.

· I remind people that this was devised by an eminent doctor and gives us a simple but clear model for how our brains are working.

Guided discussion – testing the experiment
10 minutes

I ask how we can test the experiment. (We can look at our own behaviour to see if there is any evidence that the results match reality.) I ask for examples. Examples are numerous and include always tying shoelaces the same way, morning routine, how we greet each other etc.

Evaluation of the way the brain works
15 minutes

I then ask 'What are the plus points of how it works?' and map them onto the flipchart, then 'What are the minus points?', and then 'What are the interesting points?'

This can also be done in small groups, using sheets of paper with Plus, Minus and Interesting as headings. Two minutes per heading. Points raised are then collated with the whole group.

Either way, points that are raised can be explored and expanded upon as required.

Guided discussion – social ecology
10 minutes

The next stage yields very interesting results, and starts by me going back to the logic bubble drawing and adding a body, arms and legs to the dots, which have now become heads, and saying 'If this operates at the level of personal ecology, might it also operate at the level of social ecology?'

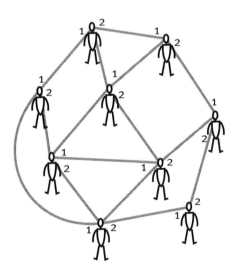

After discussing for a couple of minutes, again I ask if there are real-life examples of these patterns operating. Then I say 'Lets assume that it does operate at this level. What might be the plus points?'. I map them and then move to minus points, map them, and then move to interesting points and map them. If it hasn't come out anyway, I always push them on the minus points until someone says WAR and PREJUDICE, which are clearly major minus points!

A word of warning: practise this yourself and then with friends and family before running it on a Design Course. It can be very confusing if not done properly.

Short break
10 minutes

There are usually a few people who want to ask more questions and this is a good time to deal with them individually.

PMI (a solution)
60 minutes

Background to the tool – talk
5 minutes
- I start by drawing the PMI on a flipchart, black/white board, or biggest landscape paper possible.

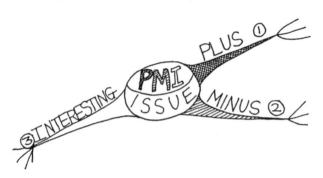

- Then give a short history of the tool: Edward De Bono developed it (Refs 2 & 3), Tony Buzan invented mind-mapping (Ref 4) and Wolf White (Ref 5) put the two together to create this Thinking Key. I tell people that this is only one of around 70 thinking tools, each of which has been developed to deal with particular thinking tasks.
- Next I introduce the rules of the tool, which are:
  - Everyone focuses on the same thinking direction at the same time ie all focus on Plus points first, then all on Minus points, then all on Interesting points.
  - No going back and forth between branches, otherwise we get back into adversarial thinking.
  - If people want to add more plus points, just go round the whole thing again.
  - Say the point as you like, but then sum it up in KEYWORDS so that it can be mind-mapped.
  - Every input is to be valued and respected: if someone gives a Plus point that another considers to be a Minus point, then the Plus point goes down, and when we get to the Minus points it can go down again! Don't dismiss the Plus point.

Practise PMIs – whole class exercise
15 minutes
- We now do a couple of quick PMIs (two minutes maximum per branch) on issues that are possibly humorous and definitely non-contentious, such as 'plasticine buses' or 'cinemas without seats'. At this stage learning the process of the tool is much more important than the content.
- Each time I start a new PMI, I ask the group what the rules are and how to draw it on the paper.

Group chosen topic – teacher directed, or in small groups
10 minutes
After the practice PMIs the group choose an issue that they would like to think about, although I will not allow topics which are likely to cause distress for any of the students. This can be a real-world issue such as their town, motorways, or a shared project they may have.

Development
10 minutes
When they have completed their PMI, I then introduce a new branch – Initiatives. Ask them to imagine that they have become the government for the day and they have no constraints of money or resources. What would they do about this issue? (I say this because people often dismiss ideas for initiatives before they have even said them, because 'they couldn't afford them'.)

   The initiatives branch is added so that the PMI evaluation leads to action, thus making it a powerful design tool.

Visualisation

5 minutes

Towards the end of the session I ask them to shut their eyes and lead a guided visualisation in which students construct and colour in the PMI in their imagination to memorise it.

Evaluation

5 minutes

I get them to evaluate the tool by doing a PMI on the PMI. I address the various points that arise to clarify the tool's use and eliminate any latent confusions. I remind students that this is only one of a set of thinking tools.

Discussion

10 minutes

Link to ethics – I ask people how using this tool may help us to put our ethics into practice.

Agenda 21 link – I ask people to discuss how this tool could be used within the Local Agenda 21 process.

Further activities

· Throughout the course, I make sure each participant has had the opportunity to do a PMI, to evaluate days, designs or particular issues that come up. It's also useful for setting home research, such as 'Help friends and family to do a PMI, with initiatives on their neighbourhood.'

· When the process for using the PMI has been established, then this can be applied to the whole range of Edward de Bono's Thinking Tools. These form a very useful addition to the current design tools.

Bibliography

1 The Happiness Purpose, De Bono, E, Penguin Books, 1990. Chapter 11 deals with the Logic Bubble
2 Teach Your Child How To Think, De Bono, E, Penguin, 1993
3 Serious Creativity
4 The Mind Map Book
5 HSDI: a Human Scale Development Initiative, 8b, Vicars Road, Leeds, LS8 5AS. (0113) 240 0349. Thinking Keys form part of the overall strategy for community development. Training, research and networking, with charts, posters and booklets also available.

Visual, Graphic and Presentation Skills

Chris Mackenzie-Davey

Why I include this session

It's an indispensable site design tool from initial survey through the design process to presentation, implementation or portfolio. 'A picture is worth a thousand words', when presenting anything from simple to complex work. It is especially important where language or literacy barriers exist.

Objective

To develop drawing and presentation skills, and raise awareness of the range of techniques and approaches available to the designer.

Learning outcomes

By the end of the session students will be able to:
· Draw simple base maps and sectional elevations.
· List different techniques and approaches for drawing and presenting design work.

Context

Mid-course, before the practical design exercises. Or introduced in small chunks throughout the course as opportunities and examples appear in the teacher's existing curriculum.

Duration

60-90 minutes

How I teach this session

Brainstorm & short presentation

45 minutes

Ask students to think of different ways of presenting designs.
· Base plans, scales, scale rulers, paper sizes, sources of information, measured plans, layers. (Note importance of accuracy at this stage.)
· Levels, contour plans, models, graphic representation, photographs, perspective sketches, axonometrics, sections.
· Zoning, sectoring, master planning, detail design.
· Presentation drawings, working drawings, construction details, planting plans.
· Presentation of complex material, flow diagrams, mind-maps, graphs.

I check to see if all the students understand the different techniques listed, explaining where necessary.

I follow this with varied examples of drawings and presentation material, taking care not to intimidate the students by showing only polished work.

One or more practical exercises in groups are great, especially if done 'live' as part of the design practicals.

Example exercise

15 minutes

Explain 'sectional elevation', and ask students to draw one for the room or space they are in.

○ Further activities

Build a model of the site, or prepare a photo report. Refer to this session again before the main design exercise.

○ Further research

Manual of Graphic Techniques for Architects, Graphic Designers & Artists, Porter, T, & Greenstreet, B, Astragal Books, 1980. How-to book for maps and plans.

 # Participative Community Planning

Judith Hanna

○ Why I include this session

Practical People Care techniques exist for drawing the local community into planning for sustainability. They embody the principle of Working With Nature, by creating opportunities for people to be part of creating the solutions to local problems, rather than working against human nature, by imposing an outsider's technical prescription which often arouses opposition and cynicism. Local people are also a huge resource. They are the ones who know their own locality, and make up its local networks and economy.

Such techniques aren't magic wands, nor simple and problem free, but they are essential for serious permaculture action beyond our own front gate.

○ Objective

To enable students to draw in useful community input and support as part of a project design process.

○ Learning outcomes

By the end of this session students will be able to:
· Have a practical understanding of what 'community' is.
· Understand why and how to engage community networks in what they are doing.
· Be aware of some techniques which are successful in drawing others' ideas into co-operative design processes.
· Feel confident to use the techniques discussed in class.
· Be aware of sources of further information and ideas.

○ Context

It follows on from Local Agenda 21. It can also follow or precede Official Systems. Transport issues are often at the top of local concerns, and exercises which link these two topics together are suggested below, read the two sessions together. Sessions which involve planning for Community Economic Initiatives or Sustainable Livelihoods can also make use of these techniques.

○ Duration

Part 1: 40 minutes
Part 2: 40 minutes

○ How I teach this session

Part 1

Starting Exercise 1. What's the local social ecology?
20 minutes
· Brainstorm onto a mind-map:
  · who makes up the community and who do we need to involve?

- what networks or organisations can be used to reach each group?
- ways of contacting them?
- interconnections?

The result should be a chart of a social ecology.

If you want to take it further, then:

Exercise 2. Prioritise key links

20 minutes

- Who are the most essential?
- What is likely to give us the best support and most useful input for the least work or expense?
- Draw up a manageable strategy.

NB: Setting some things as priorities means other things are less urgent or important. Trying to do everything all at once is a recipe for confusion, burn-out and bankruptcy. Burn-out is the number one hazard for community activists. Set limits on what you can sensibly do, and stick to them.

Part 2

Information session

40 minutes

Using a flipchart or black/white board, give an overview of the range of community planning techniques.

Start by asking if anyone has been involved in or made use of a community planning process. Ask them to describe it briefly. List contributions on the board.

Then run through techniques not mentioned by the class. The checklist below gives a basic set. Teachers and students may be able to add to these:

Checklist of techniques:

- Planning for Real: for site-based planning, using a model of the area, which is often made by local children. Notes are placed on the model to identify problems and solutions and moved around until everyone is happy with what's to happen where (Refs 1 & 2).
- Visioning: engages storytelling and imagination. Start by asking people what they would ideally like to see happen in their neighbourhood in, say, 20 years time – 'happy ever after' endings. Encourage humour, inventiveness, idealism. Next stages involve working backwards to draw up a practical action plan. This can be in the same meeting, getting people to think about 'How far would we get in five years time?', then 'What are our first steps now?' and 'What could we achieve this year?' (Ref 3).
- Conflict resolution: if there are two or more very different visions, then identify the two or more different aims. Divide into groups of people who share the same aim (eg ban all cars vs traffic needs to flow vs a bit of both). Ask each group to come up with, say, 10 actions it thinks are most needed now. Generally, each group will come up with actions it hopes may be acceptable to the opposing camp, but hasn't wasted time arguing with them. Often, some actions are suggested by both, eg more zebra crossings, cycle parking, bus lanes, exhaust pollution limits. Full meeting votes which actions to support or carry out (Ref 4).
- Wants and offers: based on LETS. Get people to brainstorm and write up on the noticeboard or post-it notes what they want the meeting to make happen. Group in main themes, eg make a kids playground, community noticeboard or newsletter. Point out that these will only happen if people put in the work needed. Ask people to write down offers of what they'll do and their name and telephone number. Give the opportunity for people to stand up and tell everyone their offer, particularly if they're looking for help or allies, or to spark off ideas. But allow people who don't want to go public to quietly hand in their offers.

- Maps and displays: a variety of interactive methods have been used. These include involving people in drawing, or sewing, parish maps, showing how they see their own neighbourhood. Flags can be planted showing what people value most (green) and what needs tackling (red). On the green flags, get them to write: what they value, how it could be improved, name and address. On the red: what's the problem, what should be done about it; how could they help, name and address.
- Networking with the groups identified in the start-up session, and especially with any Local Agenda 21 process.

Living examples

Ref 5 contains numerous examples.

Further activities

If time permits, and especially if students are working on a local project together, then actually set up and run through one of the techniques discussed. This could be as a class session, or taken out to a wider community setting.

Agenda 21 link

Chapters 10, 14(B), 24-32

Bibliography

1 The Power in our Hands: Neighbourhood based world shaking, Gibson, T, Jon Carpenter, 1997
2 Neighbourhood Initiatives Foundation, The Poplars, Lightmoor, Telford TF4 3QN. Tel: (01952) 590 777
3 Participation Works, (round-up of participatory planning processes and tool-kits), New Economics Foundation 1998
4 Sustainable Transport Solutions, Hanna, J, Permaculture Association (Britain), 1997
5 Local Agenda 21 Case Studies

Further research

Starting with a Seed... a Guide to Community Projects in the Environment

Local Agenda 21 Cookbook

New Economics Foundation, Cinnamon House, 6-8 Cole Street, London, SE1 4YH. Tel: 020 7407 7447. Email: neweconomics@gn.apc.org

⇨ Involving People

Dave Melling

○ Why I include this session

The creation of a more sustainable society can only be achieved by everyone having the opportunity to contribute on an equal basis. By being involved in the activities leading towards this they will understand what is happening and be more likely to own the actual change process. Agenda 21 makes it clear that progress can only be made when agencies work together and involve local communities in the design and development of the future.

This session also acts as a cautionary note and reminds students that the design process is fundamentally about listening. Good designs build in a process of handing over responsibility. It's about empowerment, not imposition.

○ Objective

To present the theories of participation, and provide an opportunity for students to design strategies that put these theories into practice.

○ Learning outcomes

By the end of the session, students will be able to:
· Explain the concepts of 'hierarchy of needs' and 'ladder of participation', and some of their implications.
· Develop strategies that can be used to involve people in design and decision-making.

○ Context

It can come after design methodology and land use planning, or sessions on communities.

○ Duration

Part 1: 70 minutes
Part 2: 45 minutes

○ How I teach this session

Opening
10 minutes
I always use a good ice-breaker for this session to get students involved. This can be as simple as asking people to pair up and share something good that happened that day or week.

Part 1: Theories of Participation – a starting point

Talk
10 minutes
Describe Maslow's Hierarchy of Human needs (Ref 1)

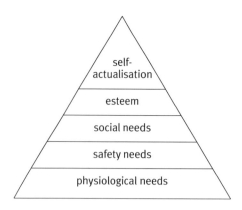

The hierachy of human needs

A H Maslow

Maslow suggests that basic human needs are arranged in loose hierachy. The more urgent needs dominate behaviour until they are satisfied and then the 'higher' needs emerge to motivate behaviour. When a need has been satisfied it no longer motivates, until deprivation occurs again. Thus if lower level needs are constantly gratified, it will be the higher level needs which tend to dominate. Usually the higher level needs are never completely satisfied unlike those of the lower levels.

The needs are arranged into five levels:

1 Psychological needs – food, sleep etc.

2 Safety needs – security, shelter and protection. These are associated with childhood and tend to emerge in adults in times of crisis or emergency.

3 Social needs – acceptance, belonging and affection.

4 Esteem needs – feeling valued, and thought well of by oneself and by others.

5 The need for self-actualisation – fulfilling or realising oneself in a growing, creative way. Achieving everything you are capable of.

Along with these there is a recognised desire to know and understand.

Maslow suggests that there are prerequisites for the satisfaction of basic needs, these include: freedom to speak, to express oneself, to defend oneself, to investigate, and to generally do as you wish.

The hierachical ordering of needs is not rigid and will certainly vary from one person to another. As a person's goals change then a different structure may appear. The movement of needs from one level to another is only a tendency not a certainty.

According to Maslow a satisfied need is not a motivator. He suggests that people are usually partially satisfied and partially unsatisfied in all of their basic needs at one and the same time. As the hierachy of needs is ascended the amount of non-satisfaction is increased.

D Melling

Thinking exercise

15 minutes discussion, 5 minutes feedback

What implications does this have for how and when people become involved in activities?

Talk

15 minutes

Describe Arnstein's Ladder of Participation with examples (Ref 2).

Thinking exercise

What are the challenges faced by society as we progress up the ladder?

Part 2: Techniques of Involvement – theory into practice

Activity

30 minutes

Take an issue such as food, alternative economy, or waste, and ask the students to devise a strategy for involving people in a programme that supports action on the issue. The group needs to outline at least eight distinct types of activities.

Examples include: forums, focus groups, questionnaires, appraisals/audits, quizzes, Planning for Real, community arts, consensus building, action planning, awards, information services, and practical projects.

○ Discussion

15 minutes

Presentation by students of their strategy.

○ Further activities

Consensus building, communications, skill sharing, visioning, managing volunteers.

○ Agenda 21 link

Chapters 3(A), 8 & 23-32.

Bibliography

1 Towards a Psychology of Being, Maslow, A, H, John Wiley, 3rd edition 1998
2 Local Authorities and community action, Warburton, D, NEST, 1993

Further research

Working in Neighbourhoods, WWF, 1997

Community Participation in Local Agenda 21, Local Government Management Board, 1994

Interactive Magazine, pub annually by Shell Better Britain Campaign

Guide to Effective Participation, Wilcox, D, Partnership Books, 1994

Consensus Building, Baines, J, The Environment Council, 1995

Our Backyard, Ed Martin Wright, Headway, 1991

Developing Sustainable Communities, A Field Workers Guide, Langford, A, Devon County Council, 1995

Local Agenda 21 Case Studies

Starting with a Seed... a Guide to Community Projects in the Environment

Short Design Exercise

Patrick Whitefield

Why I include this session

a) Some of my courses consist of an introductory weekend plus two five-day blocks. I always include a design exercise on an introductory weekend, so it becomes part of a design course run on this format. I believe a design exercise is essential on an introductory weekend. Many students say that the exercise was the time they really learned. I once omitted the design exercise, and by chance a student on that weekend attended another one taught by me, as part of his design course. He was emphatic that the second one, with a design exercise, was much better.

b) I often include it on other courses, but if there is a particularly rich supply of outdoor activities such as visits, I may not.

Objectives

To give students a brief experience of designing, and thus consolidate what they have learnt in the course so far, and to bring home to them how much of the relevant skills and knowledge they already have.

Learning outcomes

By the end of the session students should be able to:
· approach the design of their own garden with some confidence.
· have an idea of what permaculture is all about which is based on their own experience.

Context

In the case of a), on the second day of the weekend; b), towards the end of the first week.

Duration

155 minutes plus

How I teach this session

For this short exercise I divide the students into groups at random. Three to five people is about the right size. With a large number of students overall it may be necessary to go over five per group so as to have enough time for design presentations: with more than four groups the amount of time for each presentation becomes too short.

If possible I choose a domestic garden as the design site. These are always the most successful, perhaps because the relationship between people and land is more familiar to most people in this context than any other. Municipal sites are particularly difficult. A garden which hasn't had much done to it tends to be a better subject than one which is already intensively used.

If there is someone with a garden nearby who actually wants some design input this usually works well, with the actual occupant answering questions at the beginning of the process and receiving the design presentations. If there is not, but there is someone who's prepared to make their garden available, I or some other member of the course staff take the part of the inhabitant. Working with someone who doesn't want their garden redesigned can be disastrous. Either, at the questioning stage, they display so little interest that the students get dispirited; or, at the design presentation stage, they find so much wrong with the designs that the students feel demolished. Both of these scenarios have happened on courses I have taught.

Where there is no domestic garden available, I often take a piece of land and ask the students to imagine that it is someone's garden, pointing out where the imaginary house is, with its back door, kitchen window etc. This works well.

Timing

· Talk on design methods – BREDIME*	20 minutes
· Design questionnaire	15 minutes
· Designing, on site	30 minutes
· Lunch break	
· Designing, on paper	45 minutes
· Design presentations	45 minutes

If necessary the times can be somewhat shorter, but this is usually just right.

If the design site is not near the teaching site, time must be allowed for travelling.

* This is normally taught as BREDIM. I add an E for ongoing Evaluation, which I feel to be a crucial part of the process.

⇨ Design Exercises

Patrick Whitefield

The following sessions have been included together so that readers can see how the design exercise process works as a whole.

1 Choosing the Site and Getting into Groups
2 Listening to the Landscape
3 Base Map Exercise
4 Design Questionnaire
5 Main Design Exercise
6 Design Presentations & Debrief

LISTENING TO THE LANDSCAPE ⇨ BASE MAP ⇨ DESIGN QUESTIONNAIRE ⇨ MAIN DESIGN EXERCISE ⇨ PRESENTATIONS AND DEBRIEF

Remember it's an Exercise

Before all design exercises I tell the students:

· Remember this is an exercise. The main object of the exercises is for you to learn how to design. If you should also come up with a good design that's a bonus – even more so if the design actually gets implemented. But don't feel under any pressure to do so. If you make a total mess of the design but learn a lot about designing in the process, then the exercise is a success.

· Having said that, people often do come up with good designs and you may well do so too. But don't get attached to it.

I say much the same before the base map exercise: 'It's not the quality of the final map that matters, it's how much you've learnt about mapping.'

It's also important to remind anyone whose land we are using for the exercise that the designs are being prepared by students, most of whom are doing the job for the very first time. Although the designs may contain some very usable ideas, they are most unlikely to be up to a fully professional standard.

1 Choosing the Site and Getting into Groups

There may be a number of contrasting design sites from which students can choose. If there is only one it should be a comparatively large-scale site to give contrast with the Short Design Exercise. (See pages 131-132.)

I ask students to work in groups, ideally of three to five people, which they choose informally amongst themselves. I add that if anyone really wants to work on their own they can, though I don't recommend it. There is very rarely anyone who wants to, but some appreciate the freedom to feel they can.

The students remain in the same group, and work on the same design site if there is a choice, through all the exercises described below. But if an individual wishes to change group for some reason I don't object.

2 Listening to the Landscape

If the site of the Main Design Exercise is the right size for Listening to the Landscape then it is used for this exercise. Otherwise another piece of land, or part of the main design site, can be used. See pages 148-149 for a full description.

3 Base Map Exercise

Why I include this session

Mapping is fundamental to design. The base map exercise also:

· Helps to focus on the importance of the receptive part of the process.
· It is often an exercise in time management.

Objective

To give students an experience of the map-making process, and to learn about maps in an active, experiential way.

Learning outcomes

By the end of the session students should be able to:

· Explain the importance of map-making in the permaculture design process.
· Explain the different techniques which may be used in map-making.
· Select those which are appropriate for a specific task.
· Make a realistic estimate of the time needed to complete a specific map-making task.
· With further experience but without more instruction, become competent map-makers.

Context

This is the first part of the Main Design Exercise, unless Listening to the Landscape has been carried out on the design site as a whole. It is usually placed early in the second week.

It takes place in the afternoon, with the mapping information session immediately before lunch.

Duration

The whole afternoon, ie three hours. Some students may continue into the evening if they have an ambitious map and want to finish it.

How I teach this session

At the end of the morning's work I ask the students to finalise their design groups by the end of the lunch break.

If the site is small, they make the map from scratch, using the surveying skills they have learned earlier in the course. If it is large they start from an ordnance survey map.

I point out that lack of time may be a problem, and advise them to concentrate on the essentials. I also remind them that it doesn't matter if they don't finish – learning is the priority. But if they get stuck on one part of the process they should leave it and go onto the next, so as to get some experience of the whole process.

Sometimes, on a site which is difficult to map, it becomes clear that no group is going to end up with a usable map. In that case I often finish the map for them as a demonstration. If all groups are designing one site and one comes up with a much better base map than the others, this can be made available to them all for the Main Design Exercise.

Where the same site is used year after year, a good base map can be saved from one year to be used in future years, either in case no-one manages to make a good map, or as a starting point for the new course to improve on.

4 Design Questionnaire (listening to the people)

Why I include this session

Listening to the people is as important as listening to the land. I call this exercise Design Questionnaire rather than Client Interview to get away from the design consultant model. Very few, if any, of the students will go on to work as design consultants, and it's important to emphasise that permaculture design skills are very much the same whether we are working on our own place or someone else's.

Objective

To give students an experience of this part of the design process.

Learning outcomes

By the end of the session students will be able to:

a) with respect to the current design exercise -

· State the requirements of the inhabitants or users of the site (fictional or real), quantified where appropriate.
· State the resources – skills, time, money etc – which they are able to devote to it, quantified where appropriate.
· Have an intuitive feel for the design task.

b) in general -

· Approach future design questionnaires with some experience of the task.

Context

At any time before the Main Design Exercise, usually the day before. A practical session on listening skills will have been done the previous week.

Duration

40 minutes before lunch, and 90 minutes immediately after lunch.

How I teach this session

Part 1: Preparation

40 minutes

I hand out copies of an example questionnaire, and talk about it for five to 10 minutes. I then ask them to get into their design groups and compose a questionnaire tailored to this specific design job.

Part 2: Asking the questions

90 minutes

As with the short design exercise, it can be better for a member of the course staff to take the part of the inhabitant of the site rather than the real inhabitant. (See Short Design Exercise, pages 131-132.) Ideally the main teacher should be free to monitor how things are going, moving from group to group if a number of different sites are being designed. If I am taking the part of the inhabitant I occasionally allow myself to slip out of role and speak as myself if necessary. If all groups are designing the same site they can ask questions all together.

5 Main Design Exercise

Objectives

a) to give students an experience of designing.

b) to consolidate much of what they have learnt on the course.

c) to give them confidence to start designing in real life situations.

Learning outcomes

By the end of the session students will be able to:

· Approach other design tasks with some confidence.

· Discuss the pros and cons of different design approaches and methods in relation to specific design tasks.

· Be aware of some possible problems in the design process and how to overcome them.

· Have a more in-depth knowledge of those aspects of permaculture which are relevant to their design task.

Context

This is the culmination of the course. All relevant information and skills must have been covered before it starts. I put it on Days 10 and 11 of a 12-day course.

Duration

6 hours: all one afternoon and all the following morning. Doing it on two half-days rather than one solid day has a number of advantages:

· It gives a variety of activities on both days.

· It allows students to sleep on it – and perhaps dream about it – which is an important part of the design process.

· The evening provides extra time for them to work on their designs if they choose, though in practice they rarely do.

How I teach this session

At the beginning I advise the students that the observation stage of the process is not yet over, and the first part of the time should be spent in the receptive mode. As a general guide I suggest that the first half of the time, the afternoon, should be spent on the land, and the second half working on paper, with additional visits to the land as necessary to check things.

I make myself available for questions throughout the design exercise, but in practice there are very few questions. From time to time I check on how each group is getting on. If a group is getting stuck on one stage of the design, or one part of it, I may suggest they give it a certain amount of time and then leave it, or leave it for now and return to it.

Towards the end of the exercise I make sure each group knows how much time is left.

6 Design Presentations & Debrief

Objective

· To give students the satisfaction of presenting the work they have done, and give an opportunity for the tutor and others to give feedback on the design work.

· To consolidate what has been learned during the design process, including both content and methodology.

Learning outcomes

By the end of the session students will be able to:

· Explain more about both the content and methodology of permaculture design, both from feedback on their own designs and seeing those of other groups.

· Approach future design presentations with some previous experience.

· State which specific design methods they found useful in their own task and which they did not.

Context

The afternoon immediately following the main design exercise.

Duration

135-180 minutes, depending on the number of groups

How I teach this session

Part 1: Presentations

At the end of the Main Design Exercise I ask the groups to decide how they will present their designs. In most cases each member of the group presents an aspect or part of the design. But if anyone chooses not to speak that's fine.

The amount of time allotted to each group depends to some extent on the number of groups and the time available. But an initial allowance of twenty minutes plus ten minutes for questions usually works well. It's necessary to be tolerant of over-runs and have a fairly flexible programme for the rest of the afternoon to accommodate this. It's not good to hold any presentations over to the following day.

When each group has finished its presentation I ask the rest of the students if they have any comments or questions. If the actual inhabitant is present I then ask them for comments, and finally I give my own comments.

These are based on key words I jot down during the presentation. I may not mention all of the points I have noted, but select the ones which I feel are most important once I've heard the whole presentation. I always start and finish my feedback with positive comments, placing any negative comments in the middle. Some sensitivity on my part is necessary in order to avoid being either too critical, which is dispiriting, or too uncritical, which fails to help them learn as much as they could from the experience.

Part 2: Debrief

The amount of time taken by this session is variable, as different students have a different level of interest in the design process. But for some it is extremely valuable, and it should never be omitted.

The two topics for discussion are:

· Which design methods did you use, and how useful did you find them?

· How was the design process for you?

I ask each group the first question in turn, then go round again with the second question. I make sure each person has a chance to speak if they want to. If general discussion starts up I may go with it or I may ask that it be postponed until everyone has had their turn. This can lead to a discussion of designing in general, and my experience as a designer.

A Note on Design Exercises

Simon Pratt

I once a had a pattern of running a short design exercise early in the course and a longer one later. I am not now convinced of this. Where the full course is opened with an introductory weekend I can see the value and I always include a short design exercise as part of my introductory weekends. However, with a two-week course I feel the time could be better used by having a more extensive design exercise.

Robin and Skye's process, first used by Lea Harrison, Max Lindegger and Frances Lang, has much to commend it. They spread the design exercise over five afternoons in the second week, covering the topics of:

1 Observation and client interview
2 Site analysis
3 Concept plan
4 Detail plan
5 Presentation

This enables the different stages to be taken slowly and deliberately. It also means each practical session can be preceded with a brief presentation of the tools and processes available for each stage.

I would draw attention to Robin and Skye's comments on design process considerations (Ref 1).

· Choice is important. Have a range of small and large scale design exercises to choose from. Don't limit them to land-based design, they can also be social or economic.
· Team work is an important part of the process, three to five people is ideal.
· It's better to have a real land owner as 'client' wherever possible, rather than a member of the teaching staff.

Bibliography
1 The Manual for Teaching Permaculture Creatively

Observation is the fundamental skill required to do good permaculture designs. It is a skill which should be taught on Day 1, Day 2, Day 3, Day 4... right through until the end of the course. Students should not only leave the course with a range of observation skills, but also with the habit of using them.

Broad objectives for observation (over a number of sessions):
· To explore the range of observation strategies available.
· To highlight the importance of listening and its relevance to permaculture design.
· To explore options for systematic observation and provide opportunities for practice.
· To encourage non-judgemental observation as the pre-requisite for good design work.
· To introduce students to the concept of reading the landscape.
· To explain the importance of measurement, quantitative data and well-controlled experiments.

 # Listening Skills
Cathy Whitefield

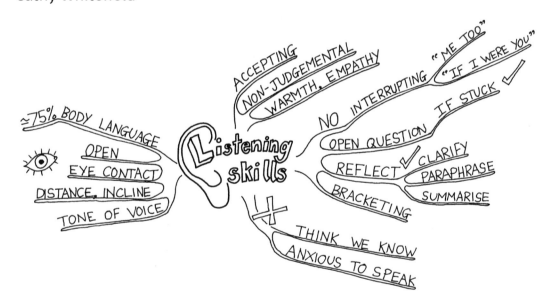

○ ## Why I include this session

Being able to listen and hear each other is the most important aspect of a relationship. It is the most important part of communication, and the one which is least valued and understood in our culture.

Listening, both to the land and to the people, is absolutely fundamental to permaculture. It's only when we can really hear the needs of the people that we can match our designs to those needs, and only designs which do meet people's needs will be implemented successfully.

○ *Objective*

To highlight the importance of listening and its relevance to permaculture design, and to give an experiential introduction to the subject.

○ *Learning outcomes*

By the end of the session students will be able to distinguish between good and bad listening, and have had an experience of both giving and receiving active listening.

○ *Context*

Usually halfway through the first week. It should be early enough in the course so that the skills learned can be used and thus consolidated later on (see 'Communities', on pages 327-328), but far enough in that students have settled down and appreciate a change from the Earth-based content of the course so far.

○ *Duration*

45 minutes

○ *How I teach this session*

Opening
1-2 minutes
I ask the students to sit comfortably, close their eyes, and listen to themselves: how they feel emotionally, eg happy, stressed etc, and physically, eg tension, aches and pains.
I then say that it's as important to listen to ourselves as it is to others.

Brainstorm
15 minutes
What makes good listening?
I comment on and clarify the results of the brainstorm, and fill in any points which have been missed. (See mind-map for check list.)

Exercise
20 minutes
Before starting, I say that some people may find it difficult and it can feel easier to have a general discussion instead of doing the exercise. I urge them not to do this: although active listening may not come easily it is very worthwhile. I also tell them not to worry if it feels unnatural. This is after all an exercise, not a 'natural' situation. Having learned listening skills it is possible to incorporate them into daily life in a more natural way.

I ask the students to get into pairs. One partner speaks for five minutes while the other actively listens, incorporating some of the listening skills from the brainstorm. The subject should not be too emotive. I normally ask them to talk on 'How the course is going for me'.

The speaker then gives constructive feedback to the listener on how they listened, and anything they could have done differently. They may also like to share their personal experience of having undivided listening attention. Many of us are unaccustomed to this and may find it difficult. (2-3 minutes.)

Then the roles are reversed. Each person thus has an uninterrupted seven to eight minutes. I tell the group when the initial five minutes is up, asking them to start giving their feedback, and also tell them when it's time to change person or complete the exercise.

Closing

Maximum of 10 minutes

When they have finished the exercise I bring the group together and ask if anyone would like to share anything with the whole group. I make sure no-one monopolises the time.

If you have not done this kind of work before, I suggest you try it on a group of friends before using it on a Design Course.

A Practical Exercise in Swapping Time

Chris Dixon

Why I include this session

I include this session for a number of reasons:

- It's a good ice breaker.
- To review course material.
- To cool students off if things get a bit heavy or excited.
- To help students get to know each other.
- As an introduction to counselling.
- It teaches listening skills, which is a form of observation.
- It teaches thinking and expressing skills.
- As preparation for making judgments that relate to design decisions, ie weighing up advantages and disadvantages of two or more courses of action.

Objective

To develop basic counselling skills.

Learning outcomes

By the end of this session students will be able to:

- Recognise the importance of developing good listening skills.
- Recognise the importance of developing good thinking skills and modes of expression.

Context

On the first day and then regularly, throughout the course. People soon become accustomed to using the technique.

Duration

20 minutes

How I teach this session

Split the group into pairs. Ask one member of each pair to raise their hand; tell them that they are going to talk first. This saves messing about deciding who is going to go first. The other person is just going to listen.

Give them a clear topic

For example, early in a course, something you enjoy doing. Later in the course, relate it to what has just been discussed; what you find most interesting about water management. Others might be, what will you do after the course? How will what you have learned effect your life? What did you like about the way this course was taught? And so on.

Timing

I tell them that the talker has two minutes to talk, whilst the other listens. Then they swap roles and do the same for another two minutes. This is given a short-hand phrase of 'two minutes each way' (2×2).

Main point

I emphasise the importance of just listening if you are the listener; you shouldn't be commenting or advising, just listening. The better you just listen, the easier the talker will find it to think and talk. The listener can prompt if the talker dries up, but silence is fine.

Development

- Results of swapping time can be fed back to the whole group where appropriate by following up with a group circle, with people speaking in turn. For example: prior to making a design decision like 'should the polytunnel go here or there?', appreciation of the course leader, or next steps.
- As students' listening skills grow, 2×2 can be extended to 5×5 or even 15×15.

⇨ *Observation*

Simon Pratt

○ *Objective*

To explore options for systematic and thoughtful observation and provide opportunities for practice.

○ *Learning outcomes*

By the end of this session students will be able to:
· Explain methods of observation.
· Carry out thematic and non-selective observations.
· Begin a 'Biotime' diary.

○ *Context*

Observation helps to integrate material learnt in the classroom. It enables students to look at things in a different way, and so more than one session is essential. Ideally a short session on observation would be held every day, in one guise or other.

○ *Duration*

Site observation: 40 minutes
Biotime: 40 minutes

○ *How I teach this session*

1 Site observation

40 minutes

After setting the objectives for the session I introduce Mollison's two sub-divisions: thematic, and non-selective or child-like. I ask students to observe part of the site using these two methods, by either giving them a theme to follow (such as trees) or letting them choose. This is followed by a short feedback session where students can share their observations of the site and the process. Other strategies for observation can also be discussed at this stage.

2 Biotime

40 minutes

I introduce the concept of biotime or phenomenological time, and recommend deliberate observation and record keeping over a number of years. Emphasise seasonal variations. Ask students to consider what they may wish to observe over an extended period of time. What is most relevant to their situation?

Encourage observation and direct experience of the physical characteristics of landscape: wind, water, slope, sun, moisture, soil, vegetation. Invite people to think 'all senses active at once!'
Handouts
I use Andy Langford's year on year calendar as an example (see below).

Activites	Date	Year
	Moon	Bio d
	Temp range	Sum
	Cloud cover	
	Precipitation	
	Sun rise	
	Sun set	
	Wind direction	
	Wind speed	
	Wild edibles in season	
Chosen species include:	Trees – buds, blossoms, leaves	
	Rivers/streams – levels	

○ Further activities

- Students walk with their eyes shut and with a guide (and barefoot if they want to), through the landscape in order to fully open the other senses.

- Using earth from different places, such as woodland, garden and degraded land, get students to smell each and determine which sample came from where.

○ Further research

Sharing Nature with Children, Cornell, J, Dawn Publications, 1979. Also by the same author are Journey to the Heart of Nature and Sharing the Joy of Nature. Both of which contain observation games and exercises.

Observation Exercise

Chris Dixon

Objective

To encourage non-judgemental observation as the prerequisite for good design work.

Learning outcomes

By the end of this session students will be able to:
· Understand the importance of good observation.
· Identify and remove assumptions from observations.
· Distinguish between random and thematic observations.
· Understand the cyclic process of: observation – speculation – hypothesis – experiment – further observation.

Context

Useful as a more practical break between head work. Good for getting small groups going. Various methods of feedback can be used, eg speaking in turn, shouting out.

Duration

60 minutes, including two 20 minute practicals; or two longer sessions, 60 minutes each, spaced during the course.

How I teach this session

Introductory talk – interactive

10 minutes
· The importance of observation in relation to site and system analysis.
· The importance of removing prejudice or assumptions.
· Working With Nature – we've got to know what she's up to first by observation.
· The distinction between random and thematic observation.
· Observation is broadened to include the other senses – sound, smell, taste, touch.

Practical 1

Small groups – random observations
15-20 minutes
I send the students off in small groups to generate sets of 5-10 random observations.

Whole group – collation and observations
20 minutes
The groups return and I collate their observations on the board.

Observations have to be examined as to whether they contain implicit assumptions. Only pure observations are allowed, and assumptions get stripped out, eg 'there are rabbit droppings on the path' becomes 'there are droppings on the path'.

The observations may generate speculations, eg 'maybe the droppings are rabbit droppings?' They may also begin to group themselves, as some will be obviously related.

Practical 2

Small groups – thematic observations
15-20 minutes
I send them off in small groups to make thematic observations, eg droppings, or rabbits.

Whole group – collation and observations
20 minutes
The observations are collated in the same way as Practical 1, and speculations and hypotheses can be generated from them.

Further observations or small scale experiments can then be designed and conducted. The results of the experiments can be observed, which may lead to conclusions or to further speculation, observation and so on.

For example, observations that conifer needles under plantations appear to be aligned in some way may lead to the speculation that run-off is moving them. Small areas, say one metre square, can be marked with coloured wool or similar. On some areas the needles can be moved through right angles, on others stirred randomly, on others larger twigs or other objects might be positioned. The areas can be checked after rainfall. What has happened, if anything? How might we use this effect in a design? In plantings? On a larger scale?

This sort of 'playing' works particularly well on a design course spread over several weeks or months.

Further activities

An observation book, sheet or board can be maintained throughout the course. Course members are encouraged to note random observations in or on it. The observations can be speculated upon and grouped a number of times during the course. People Care stuff tends to come out of this one.

Basic ideas for this came from Bill Mollison, Ragman's Lane Course, Spring 1991.

Listening to the Landscape

Patrick Whitefield

Why I include this session

This is an observation exercise. In it students use the faculties of intuition, feeling and imagination as well as intellect. I include it because:

a) observation is the most important part of permaculture design, and we need to practice it.

b) in our culture, which emphasises the intellect over all else, we need to develop our other, right-brain, faculties.

I am indebted to Christopher Day, architect, of Crymych in Pembrokeshire, who taught me this exercise.

Objective

To enhance students' ability to observe.

Learning outcomes

By the end of the session students should have had:

· An experience of quiet observation which most of us never have.
· An opportunity for insights into themselves, how they observe and how they relate to the landscape.
· Some insight into the unique nature of the specific landscape they have been observing.
· An opportunity to experience their relationship to it.

Context

Not too early in the course, as the state of mind needed for this exercise is very different from our day-to-day one and people need some time to settle into the course atmosphere first. But before the Main Design Process (Base Map, Design Questionnaire and Main Design Exercise).

○ ## *Duration*

90 minutes

○ ## *How I teach this session*

Site

An area between a half and 5 acres is suitable. Big enough to take all the students without them feeling crowded by each other; small enough for them to get round most of it and to hear my signals. The boundaries must be distinct and easily understood. If the main design site is a suitable size it is ideal to use this.

Instructions

10 minutes

"I would like you to observe this area in four different modes. There is 10 minutes for each mode. I will blow a whistle when each 10 minutes is up. The modes are:

· Intuitive: your first impression, unrepeatable; walk the boundary (don't worry if you only get part way round), observe relationships with neighbouring land.
· Objective: the rational approach in which we try to place no values; different aspects of the site are observed systematically, eg soil, micro-climate, water etc; for the purposes of this exercise, record all the plants in the area – if you don't know all their names, describe them.
· Imaginative: imagine how this place was before humans ever started to influence it, and how it would change if we stopped influencing it now.
· Subjective: go to whichever part of the site your body draws you to (not always the most pleasant part); if you feel moved to, express yourself in any non-rational way, eg sing, write a poem, dance, or make a sculpture.

"Please work as individuals, and in silence. You may take notes or not as you choose. At the end of the time we will come together and discuss how it was.

"This is an exercise, to give you a taste of observing in these modes. In practice you would do this over a much longer period, ideally on and off for a whole year or more.

"It's important to remember that in this exercise we're not asking 'What can I do with this land?' but 'What is this land saying to me?' It's about the needs of the land, not of the people. That comes later in the design process."

Walk to site

10 minutes

Exercise

40 minutes (not necessarily exactly 10 minutes for each mode)

Discussion

20 minutes

I invite people to discuss:

· The process, and what they learned about how they observe.
· What they learned about this piece of land.

Sometimes I start them off discussing in small groups then bring them all together. Other times I do one or the other.

Walk back

10 minutes

Reading the Landscape

Patrick Whitefield

Why I include this session

By reading the landscape I mean looking at the ecological processes which are going on around us and understanding them. The ability to do this, and the habit of doing so, are important attributes of a good permaculture designer, whether they be a consultant designer or someone working to implement permaculture in one place.

Objective

To introduce students to the concept of reading the landscape and to some specific interactions which can be seen.

Learning outcomes

By the end of the session students will be able to:
· Start developing the habit of reading the landscape.

Context

This is not something which needs to be covered before the main design exercise, so I often place it late in the course. It can also be scheduled as an evening activity, as it's more relaxed than many other topics, as most students don't feel the need to take notes or remember specific facts.

Duration

30-45 minutes

How I teach this session

A slide show of 35-40 slides, illustrating:
· Soil indicator plants; how the same plant can indicate different things.
· Plant condition as a soil indicator.
· Plants as micro-climate indicators.
· The interaction of soil, climatic and biotic factors.
· The influence of grazing on moorland, woodland and trees.
· The influence of the sea.
· Plant-plant interactions.
· Degeneration and regeneration of woodland.
· Historical aspects.

All the slides are taken by myself. I am very happy to share them with other teachers, although it can be difficult to answer questions about pictures you didn't take yourself. I would recommend any teacher of permaculture to develop the habit of reading the landscape. Always carry a camera loaded with slide film wherever you go.

Living examples

All around you!

Further activities

This links with the Listening to the 'Landscape exercise' (see pages 148-149), which is experiential, whereas this is knowledge-based.

Further research

The History of the Countryside

The Ecology of Urban Habitats

The Wild Flowers of Britain and Northern Europe

Trees and Bushes in Wood and Hedgerow, Vedel, H & Lange, J, Methuen, 1958

➡️ *Everything is a Gift (Wild Food Forage)*

Graham Bell

○ *Why I include this session*

Most of us have the impression that nothing in life is free. This session offers a contradiction to that by showing how there is an abundance all around us.

○ *Objective*

To demonstrate the principles of gathering wild food and help students to identify for themselves what further learning is required to have detailed ability.

○ *Learning outcomes*

By the end of this session students will be able to:
· Identify a limited range of edible wild plants.
· Identify the limits of their knowledge and know how to expand it.

○ *Context*

Can be used on the first day (see 'The First Day', pages 12-19). This is a good exercise to use in the graveyard shift after lunch, or placed before a break, with feedback afterwards.

○ *Duration*

45 minutes

○ *How I teach this session*

Practical
Gathering edible plants.

Main points
· Learning the value of wild plants has a number of benefits, including:
  · free food
  · healthy eating
  · appreciation of diversity
  · contact with nature
· Take no more than you need.
· Leave protected and rare species.
· For safety, if in doubt DO NOT EAT.

Do not let anyone inexperienced under your care (eg children, trainees) gather unsupervised. Learn with a knowledgeable person.

○   ## *Further activities*

Wild food can be picked and used as part of the course menu, eg nettle soup.

○   ## *Further research*

Food for Free
The Permaculture Garden
Wild Food, Phillips, R, Pan Books, 1983

Everything is a Gift (2)

Graham Bell

○ **Objective**

To provide an opportunity for students to find and state the value of unused assets.

○ **Learning outcomes**

By the end of the session students will be able to look at situations with limited resources and still find valuable opportunities.

○ **Context**

Can be used on the first day (see 'The First Day', pages 12-19). This is a good exercise to use in the graveyard shift after lunch, or placed before a break, with feedback afterwards.

○ **Duration**

45 minutes

○ **How I teach this session**

Directed observational walk

I give the following instructions: "In groups of three walk for 30 minutes and note 10 unused assets you see. Agree one use for each asset."
I provide a sheet to be filled in.

An unused asset	A use
1 pile of cardboard	mulch, insulation
2 old tyres	retaining wall, for plants
3	
4	
5	
6	

Feedback – whole group

15 minutes

Give each participant an opportunity to give an example of an unused asset, and a use that was found by the group. Ask for further observations about the exercise itself, and what may have been learnt.

Doing Experiments and Collecting Data

Peter Harper

Why I include this session

At this stage in the growth of the sustainability movement it's more powerful to generate widely usable results than to try to be ever-so-green ourselves. There are not enough of us to make a significant difference to the ecological impact of the human population, yet there is a real need for reliable information on what works and what doesn't. One of the best ways of finding out is to monitor our own lives and see how effective our green practices are.

Experimental methods need to be taught on Design Courses, because improperly conducted experiments can actually lead us astray and waste everybody's time. Also, experimenting is a very effective learning tool, and can consolidate much of the material taught on the course.

Objective

To explain the importance of measurement, quantitative data and well-controlled experiments in permaculture.

Learning outcomes

By the end of the session students will be able to:
· Explain the significance of controls, confounding factors and statistical probability tests.
· Design a well-controlled experiment.

Duration

30 minutes

How I teach this session

Opening talk
10 minutes
· Even keeping very simple records is worthwhile:
  · is the size of your bin-bag going down on account of more recycling and composting?
  · how is commuting time affected by different modes of transport?
· Going further, there is a lot of scope for those with the right temperament to do serious and useful quantitative research in their houses, gardens and in the community. If many people all do a bit, it can add up to a tremendous amount.

Presentation of an example
I use one from my own experience, to illustrate the general points budding researchers need to be aware of.

General point	My example
Decide on a long term aim	To measure material flows through my house and garden system, with a view to improved design of houses, gardens and lifestyles
Focus on a manageable amount of resarch at a time	Currently, the flow of biological wastes and plant nutrients, eg urine
List all the unanswered questions about your specific subject	Urine: can it be used direct on edible crops? Which crops? Diluted or neat etc etc
Choose one of them	Can it be used on onions?
Set up an experiment with a control	One row of onions receives urine, another doesn't. The latter is known as the control, and enables me to see what effect the urine has
You may also need to consider a location	I placed a third row of onions between the treated row and the control row to catch any urine which might seep sideways
Run the experiment	Once a week I gave the experimental onions a watering can ful of urine diluted 1:20 with tap water, and the other two rows the same quantity of plain tap water
Record the results	Yields: onions with urine 90.3g intermediate row 61.1g control row 33.9g [I also recorded the weights of individual bulbs – necessary for doing the statistical tests]
Confounding factors, ie could the observed differences be due to something other than factor under investigation	Unlikely in this case, but hard to refute. Eg there could have been a difference in past manuring. I could have got round it by dividing the plot not into two but four, and splitting the two treatments diagonally like a heraldic shield. This is normal research practice
Interpret the results. There are statisitcal tests to tell you the odds of your results happening by chance. Most computers have them	Unusually, this is such a starkly positive result that probability tests are hardly necessary. But the statistical test told me that the odds against my results happening by chance are more than a thousand to one.

○ **Agenda 21 link**

Chapter 14(L) calls for observation of the effects of UV radiation on plants and animals. Throughout Agenda 21 many such examples of research and observation requirements crop up. Not all of which require large budgets or complicated equipment. (See also 11 D.)

○ **Further research**

An Introduction to Scientific Method, Wilson, EB, Dover Publications, 1991
Experimental Designs, Cochran, W, Wiley Classics, 1992
The Art of Science: A practical guide to experiments, observation and handling data, Carr, J, Hightext, 1993

Mechanical Tools – Helpful or Harmful?

Phil Corbett

Why I include this session

I feel it is important that people look at tools in a 'what's really going on here?' frame of mind, rather than assuming that a tool is automatically useful because it is a tool.

Objective

To extend the skills of observation and explore the issues around using mechanical tools.

Learning outcomes

By the end of the session students will be able to:
· Have an awareness of issues around tools use.
· Use tools that are appropriate to the job.
· Begin to re-examine common activities to find assumptions that may not be useful.

A note of caution...

Permaculture courses leave people feeling enthused and keen to start healing, protecting and enhancing our natural resources. Many with little experience of working on the land feel empowered to make a start, often wishing to create intensive Zone 1 gardens to begin food production. The amount of work needed to clear what appears to be a small area can prove daunting, as can finding, moving and laying down sufficient mulch materials. A perception of slow progress can lead to turning to more conventional tools of land management, rotavators, strimmers etc, which while seeming to finish the job are actually cosmetic or destructive in effect, and will need further work to make good the damage.

For some the joy of machinery use may compete with environmental considerations. The larger and more powerful the tool the greater is the tendency to alienate and insulate the user from what is actually being done. Soil is the ultimate small scale system where billions of microscopic elements create the 'skin' we depend on. We cannot get close enough to it to 'know' it even using our bare hands. Any further distancing of our attention reduces our understanding even more.

Context

This topic may start a lively debate with tool lovers. It is often a matter of 'hints and tips' leading to casual discussion rather than formal teaching, and is appropriate during the planning of practical work where the particular machines might be used.

There are many subjects where mechanical tools may be discussed:
· Erosion, tillage, mulching or garden making, eg rotavators.
· Tree care, eg lawnmowers, strimmers.
· Site maintenance, eg lawnmowers, strimmers.
· The layout/installation phase, eg wheelbarrow routes.
· he energy aspect of machine construction and fuel use.

○ *Duration*

A few minutes or a recurring debate.

○ *How I teach this session*

It is crucial that this issue is explored when examples are to hand, and the opportunity to make practical comparisons is available. It needs to be hands-on or it loses impact.

Practical comparison work

In a session of, say, nettle harvesting, have a person with a strimmer (ear guards, gloves, ankle boots) and a person using their hands (long sleeves, gloves), both remove nettles for mulch or compost. Each has a barrow.

· Who is quicker to fill the barrow?
· What did each notice about the nettles and nettlebed?
· What wildlife was on or amongst the nettles?
· What was each aware of in the environment around them while working?

On appropriate sites look for:

· Compaction and erosion by wheels – easy ways to prevent or minimise this?
· Examples of lawnmower and strimmer damage to trees – show the different types of scarring.
· Signs of healing tissue covering wounds.
· Signs of repeated damage in the same part of the tree.
· Signs of canker and other rots moving in.

Public parks are likely places to look, also trees close to car parking show similar damage.

Main points

· Rotavators are worse than ploughs or spades as they leave the soil almost fluffy in texture and liable to erosion by wind or water. They also kill worms, create impermeable pans and multiply weed roots.
· Most motorised tools are designed to work against nature and are polluting.
· Some are useful as harvesting aids but they have the power to be very damaging.
· Many jobs can be done more quickly, comfortably and safely with hand tools.
· Hand tools (and hands!) allow greater contact with the natural world and opportunities for learning.
· Biological solutions are usually slower, eg soil reconditioning, but preferable where haste is not essential.
· Tools often confer messages about status.
· Jobs where it is necessary to get dirty, or onto hands and knees, are often given low status.

○ *Link to principles*

· Observation: attention, awareness.
· Minimum effort for maximum effect: do-nothing.

○ *Living examples*

Dying examples everywhere.

Broad objectives for Agenda 21 sessions:

· To provide information about Agenda 21, what levels it operates on, and its implications.
· To demonstrate how Agenda 21 can be used to develop projects with wide support.
· To explain why local action is key to its success.

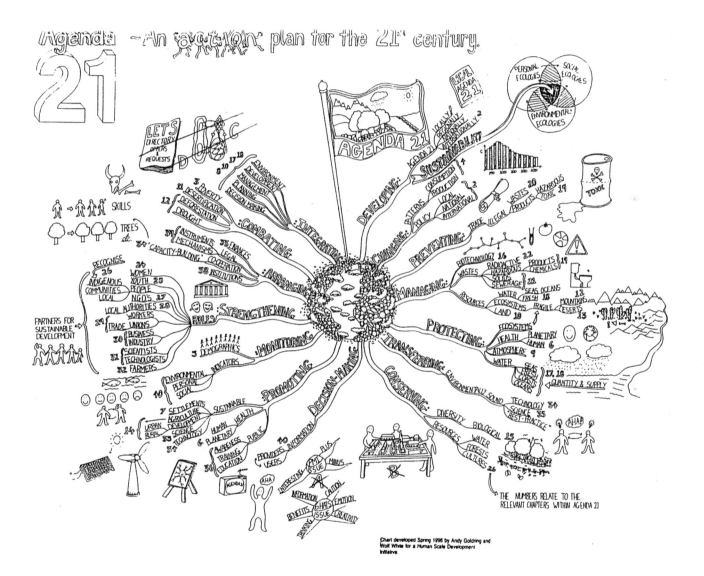

Agenda 21 – Key to Chapter Numbers

This has been included so that the reader can find out what the Agenda 21 links given in the sessions refer to. For a more detailed look at Agenda 21, see Earth Summit '92, Quarrie, J (Ed), Regency Press Corporation, 1992.

Chapter 1   Preamble

Section I   Social and Economic Dimensions
Chapter 2   International co-operation to accelerate sustainable development in developing countries and related
                 domestic policies
Chapter 3   Combating poverty
Chapter 4   Changing consumption patterns
Chapter 5   Demographic dynamics and sustainability
Chapter 6   Protecting and promoting human health
Chapter 7   Promoting sustainable human settlement development
Chapter 8   Integrating environment and development in decision-making

Section II   Conservation and Management of Resources for Development
Chapter 9   Protecting the atmosphere
Chapter 10   Integrated approach to the planning and management of land resources
Chapter 11   Combating deforestation
Chapter 12   Managing fragile ecosystems: combating deforestation and drought
Chapter 13   Managing fragile ecosystems: sustainable mountain development
Chapter 14   Promoting sustainable agriculture and rural development
Chapter 15   Conservation of biological diversity
Chapter 16   Environmentally sound management of biotechnology
Chapter 17   Protection of the oceans, all kinds of seas, including enclosed and semi-enclosed areas, and coastal areas
                 and the protection, rational use and development of their living resources
Chapter 18   Protection of the quality and supply of freshwater resources: application of integrated approaches to the
                 development and use of water resources
Chapter 19   Environmentally sound management of toxic chemicals, including prevention of illegal international traffic
                 in toxic and dangerous products
Chapter 20   Environmentally sound management of hazardous wastes, including prevention of illegal traffic in
                 hazardous wastes
Chapter 21   Environmentally sound management of solid wastes and sewage related issues
Chapter 22   Safe and environmentally sound management of radioactive wastes

Section III   Strengthening the Role of Major Groups
Chapter 23   Preamble
Chapter 24   Global action for women towards sustainable and equitable development
Chapter 25   Children and youth in sustainable development
Chapter 26   Recognising and strengthening the role of indigenous people and their communities
Chapter 27   Strengthening the role of non-governmental organisations: partners for sustainable development
Chapter 28   Local authorities' initiatives in support of Agenda 21
Chapter 29   Strengthening the role of workers and their trade unions
Chapter 30   Strengthening the role of business and industry
Chapter 31   Scientific and technological community
Chapter 32   Strengthening the role of farmers

Section IV   Means of Implementation
Chapter 33   Financial resources and mechanisms
Chapter 34   Transfer of environmentally sound technology, co-operation and capacity building
Chapter 35   Science for sustainable development
Chapter 36   Promoting education, public awareness and training
Chapter 37   National mechanisms and international co-operation for capacity building
Chapter 38   International institutional co-operation for capacity building
Chapter 39   International legal instruments and mechanisms
Chapter 40   Information for decision-making

NB: Each chapter lists the basis for action, objectives, activities and means of implementation.

Permaculture and Local Agenda 21

Jamie Saunders

Why I include this session

Agenda 21 is a mainstream response to the pressures of sustainability and as such it is connected to existing, diverse political and bureaucratic processes. Green issues have often been perceived as 'alternative', however, recent activity has seen the gradual adoption and integration of sustainability policies and practices by governments and businesses.

Permaculture supports the practical development of sustainable solutions, which by their very nature support Agenda 21. Permaculture can also support the design of Local Agenda 21 (LA21) processes and frameworks at a local level and can help in setting out the aspirations and visions of a more sustainable future.

Objective

To raise awareness of Agenda 21, and identify the main design challenges and opportunities of Agenda 21 at a local and practical level.

Learning outcomes

By the end of the session students will be able to:
· Generally understand the international, European, national, regional and local context of the adoption of the principles of sustainable development.
· Explain the challenges and opportunities of LA21.
· Explain how initiatives are designed using permaculture support and implement the objectives of LA21.
· Develop designs which take account of organisational and government backing for sustainability, connect to LA21 networks as appropriate and act as case studies for other sites and projects elsewhere.

Context

It varies, but I usually place it early in the course.

Duration

60 minutes

How I teach this session

Agenda 21 is like the ethics and principles of permaculture, something that I try to embed into the whole Design Course.

Information provision, interspersed with questions and small group discussions
60 minutes
Five overheads/handouts are used, and these are:
· Global to local model.

· Three circle model of sustainable development.
· Principles of sustainability.
· Thinking about business.
· Characteristics of a sustainable society.

Main points

Permaculture offers me three things:

· A framework for the 'conscious design of sustainable systems', which the mainstream needs to become more aware of.
· The potential for land, community and bioregional visions or longer-term designs. Again the mainstream requires a clearer idea of the impact of short-term 'bad' design.
· The opportunity to allow networks of trained people to support each other and others to make a significant impact of the implementation of Agenda 21 policies at local, regional, national or other levels.

Overhead 1: Global to local model

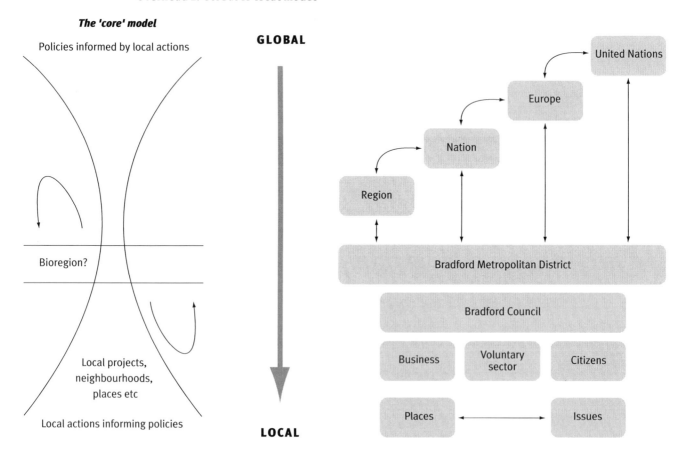

This global to local model provides a framework to discuss the various levels and sectors of society that are starting to adopt sustainability into their operations. The Agenda 21 statement that two-thirds of the challenges facing the planet must be tackled locally paves the way for Local Agenda 21. This then leads to identifying how practical permaculture initiatives can be linked to the LA21 process.

The global to local model can be drawn to match the core model (Ref 1, p73). Flows of information, such as good practice or policy recommendations, can flow from top-down to bottom-up and vice versa. The links between different levels of government, businesses and informal networks are key in enhancing the adoption of sustainability across all sectors of society.

The core model can be developed on a flipchart to show the layering of activities and levels of scale, and their impact on local conditions and projects. At this stage in the session, a sample of the levels of activity are put forward:

· United Nations – Commission on Sustainable Development (CSD)

　World Association Council Local Authorities/International Union of LAs

　International Council of Local Environmental Initiatives (ICLEI)

· European Union – Parliament and Commission

　European Sustainable Cities and Towns Campaign

· National Government – Members of Parliament and Civil Service

　(especially Department of Environment, Transport and the Regions (DETR)

　National Sustainable Development Roundtable

　Local Government Association/Improvement and Development Agency (IDA) – formely the Local

　Government Management Board – National LA21 Steering Group

　United Nations Environment & Development UK Committee

· 'Regional' Government

　Regional Assemblies/{Regional Development Agencies}/Regional Planning Conferences/other

　regional networks

· Political/Admin. boundary

　· Local Authorities (Metropolitan/County/District)

　· Local Congresses/Development agencies etc

　· LA21 networks

If done early on in a Design Course, I make an explicit link between the natural system model of ecological succession from bare earth to woodland and its lessons for LA21. For example, the role of weeds and pioneers is critical, as individuals need to lay the foundations, both in ensuring Agenda 21 is established locally and in making sure that the process is well-designed and sustainable. There is plenty of evidence of LA21 being adopted by Local Authorities with little participation, and with minimal understanding of the full implication of sustainability in terms of radical change to the status quo. Therefore 'weeds and pioneers' are needed to make appropriate interventions into the system for it to gain more widespread support.

Overhead 2: Three circle model of sustainable development

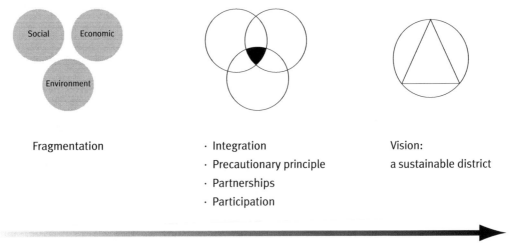

Fragmentation

· Integration
· Precautionary principle
· Partnerships
· Participation

Vision:
a sustainable district

TIME

This illustrates how the policies of Agenda 21 now explicitly link social, economic and environmental concepts into an integrated model of sustainable development. This challenges us to engage with the mainstream, to understand how organisations and governance works.

Permaculture can provide models for the development of LA21 and in ways which link these three concepts. For instance, the use of the succession model can illustrate local sustainability in economic development. In most cases local priorities are geared towards job creation 'at any cost'. Often the focus is on inward investment to bring in the factory with 1000 jobs, rather than cultivating a more diverse and smaller scale cumulative approach to the development of local communities and livelihoods.

Principles of sustainability

· Long termism and vision

· Interconnectedness

· Precautionary principle

· Participation and partnership

· Social inclusion and equality

Bradford Metropolitan District Local Agenda 21 Community Plan 1997–2000

Overhead 3: Principles of sustainability

This provides the core principles of Agenda 21 and mainstream work on sustainable development. This is where there is explicit overlap between permaculture principles and those starting to be adopted elsewhere. A useful comparison can be made.

One of the main features of LA21 is that it is enabling individuals who have done a Design Course to become active locally in sustainable development.

It also opens the door for permaculturally designed projects to act as best practice in Agenda 21, as the quality of well-designed projects is self-evident to the mainstream. For instance, the development of land-based projects, designed with permaculture, is being supported, both financially and in kind, with regeneration funds now being more readily accessed and supported. See Living examples below for examples of projects receiving local authority and other support in their establishment and development.

Thinking about business

· Environmental industries offer growing business opportunities

· Competitiveness

· Innovation in products and processes

· Minimise inputs

· Waste as a resource

· Total quality management

Bradford Metropolitan District Local Agenda 21 Community Plan 1997–2000

Overhead 4: Thinking about business

This sets out the issues surrounding language. If we can unlock jargon and find the right hooks then we can build bridges across traditional boundaries. Stake-holder responsibility, total quality management, innovation, wise resource use, import substitution and so on become areas for design strategies and solutions. This does not accept that the current economic and political systems do not need to change, but rather shows where doors might be opened for mutual support.

Overhead 5/6: Characteristics of a sustainable society

I close the session by setting out a framework for a sustainable society, as defined by the Department of Environment, Transport and the Regions (DETR), the Local Government Association and the Local Government Management Board in (now the IDA) 1998 (Ref 2). This helps bring the aspirations of Agenda 21 and permaculture together to identify common ground and project opportunities.

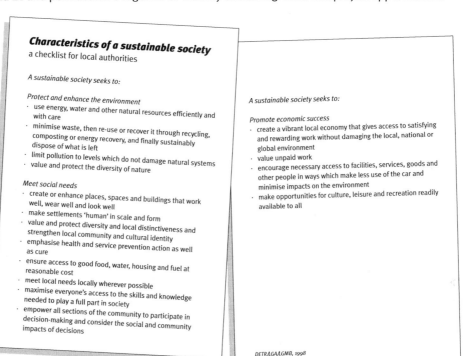

Characteristics of a sustainable society
a checklist for local authorities

A sustainable society seeks to:

Protect and enhance the environment
· use energy, water and other natural resources efficiently and with care
· minimise waste, then re-use or recover it through recycling, composting or energy recovery, and finally sustainably dispose of what is left
· limit pollution to levels which do not damage natural systems
· value and protect the diversity of nature

Meet social needs
· create or enhance places, spaces and buildings that work well, wear well and look well
· make settlements 'human' in scale and form
· value and protect diversity and local distinctiveness and strengthen local community and cultural identity
· emphasise health and service prevention action as well as cure
· ensure access to good food, water, housing and fuel at reasonable cost
· meet local needs locally wherever possible
· maximise everyone's access to the skills and knowledge needed to play a full part in society
· empower all sections of the community to participate in decision-making and consider the social and community impacts of decisions

A sustainable society seeks to:

Promote economic success
· create a vibrant local economy that gives access to satisfying and rewarding work without damaging the local, national or global environment
· value unpaid work
· encourage necessary access to facilities, services, goods and other people in ways which make less use of the car and minimise impacts on the environment
· make opportunities for culture, leisure and recreation readily available to all

DETR,LGA,LGMB, 1998

Handouts

· Copy of Overheads 1-6.
· Background material from the Improvement and Development Agency (IDA) – formely the Local Government Management Board, the lead organisation on LA21 in the UK (Ref 3):
  · Policy Initiatives for Sustainable Development/The Agenda of Agenda 21
  · 6 Steps in the Process. Source: LA21 Principles and Process: A Step by Step Guide, LGMB – now IDA (1994). (Free guidance notes on a wide range of issues as well as a number of other leaflets are also available from their publications department)
· Reading list and related Internet addresses.

There is also a wealth of material from a wide range of organisations, including WWF, and the Shell Better Britain Campaign.

○ *Link to ethics and principles*

Permaculture offers support to Agenda 21 by enabling people to have a clear appreciation of mature ethical behaviour and our role as citizens within society. The ethics and principles can also inform the development of projects and the development of LA21 processes locally. A basic definition of permaculture as 'applied common sense' is a helpful starting point for work on LA21. Permaculture allows us to take the 'best of the old and the best of the new' and synthesise them into 'the conscious design of sustainable systems' and therefore make a significant contribution to the development of a 'perma-nent culture'.

○ *Living examples*

Springfield Community Garden, Stirling Crescent, Holmewood, Bradford, BD4. Tel: (01274) 753924. See also Issue 5, Permaculture Magazine, Permanent Publications, 1994.
Drift Project, John Marley Centre, Muscott Grove, Whickham View, Scotswood, Newcastle-Upon-Tyne, NE15 6TT. Tel: 0191-200 4726 / 4735. Details in The Permaculture Plot (Ref 4).

The adoption of permaculture training is becoming more commonplace as a supported element of Local Agenda 21 in the UK. For example, Design Courses are being supported and subsidised by Bradford Council as part of its community development work for sustainability.

Bradford District's LA21 process is in itself being designed and developed, in the context of 'rolling permaculture'.

○ *Further activities*

· Greater public and organisational involvement in decision-making and implementation of sustainability and LA21. Get active and see how it goes.
· Make the links from your projects to LA21 and its regional, national and international context. This may help access resources, for instance government regeneration funds.
· See section 'Making Connections' or 'What's My Role in Life?' (pages 312-315) for personal development and livelihood opportunities.
· Ecological Footprint work (Ref 5).
· Sustainability Indicators.
· Bioregional planning (Ref 6).
· Invite your local Agenda 21 officer to speak on the course.

○ *Bibliography*

1 Permaculture: A Designers' Manual, p73
2 Sustainable Local Communities for the Twenty First Century, DETR/LGA/LGMB, 1998
3 Improving and Devopment Agency (IDA) formely the Local Government Management Board (LGMB) – the lead UK organisation on LA21. Sustainable Development Unit, LGMB, Layden House, 76-86 Turnmill Street, London EC1M 5QU. Tel: 020 7296 6599. Fax: 020 7296 6666.
· A Framework for Local Sustainability, 1994, RefLGo110
· UK Local Government Declaration on Sustainable Development, 1993
· LA21 Principles and Process A Step by Step Guide, 1994, Ref LGo116
· Local Agenda 21 Strategy Cookbook, Menu, Recipes, Ingredients, 1996
· LA21 in the UK The First Five Years, 1997, 4
4 The Permaculture Plot
5 Our Ecological Footprint
6 Boundaries of Home, Mapping for Local Empowerment, ed Aberley, D, New Society Publishers, 1993

○ ***Further research***

Agenda 21: The United Nations Conference on Environment and Development, 1992. See also outcomes from Earth Summit II Conference June 1997 and preparation for Earth Summit III

Rescue Mission Planet Earth: A Childrens' Guide to Agenda 21, Peace Child International, 1994.

Opportunities for Change, UK Government, DETR, 1998

Modern Local Government: In touch with the people, The Local Government White Paper, UK Government, 1998

The Way Forward: Beyond Agenda 21, ed Dodds, F, United Nations Env. & Dev UK, 1997

Futures by Design: The Practice of Ecological Planning, ed Aberley, D, New Society Publishers, 1994. A useful overview of a wide ranging view of progress on sustainability thinking and design

See also, Permaculture Magazine, Issues 7 & 9, Permanent Publications, 1995

UK21 c/o Tea Warehouse, 10a Lant Street, London, SE1 1QR. Tel: 020 7407 8585. Fax: 020 7407 9555 Email: pip.ltd@easynet.co.uk

Acting on Our Shared Concerns – Agenda 21

Andrew Goldring

Objective

To provide a basic background and introduction to Agenda 21. To show that the Agenda 21 action plan reflects many of our own personal and community concerns, and that it provides a big opportunity for local people to make beneficial changes in their community in partnership with local government and other organisations.

Learning outcomes

By the end of this session, students will be able to:
· Explain the background to Agenda 21.
· Describe links between local concerns and Agenda 21.
· Have confidence that new partnerships and projects are possible by using the 'common language' of Agenda 21.
· Explain why local changes are essential if we are to solve global problems.

Context

I teach this session early in the course. This enables me to make reference to specific links between permaculture theory and practice and the Agenda 21 document throughout the course.

Duration

45 minutes

How I teach this session

Opening exercise

15 minutes

"Get into pairs and spend 6 minutes exploring concerns you have for your local community and environment. Do this as a '3x3 Think and Listen'." (See pages 139-141 and 142-143 for details.)
After five minutes I pair up pairs, ie, create groups of four people.
"In your new groups see what overlap exists between the points you raised in pairs. You can add new points if they come up."

Whole group collation

10 minutes

I ask each group to put forward a point in turn, asking whether this point has come up in any other groups. If the point has come up more than once, I tick it on the flipchart to show that it is a shared concern. This continues until all points have been collated. Points can be clarified and developed at this stage.

Interactive talk

20 minutes

I start by asking whether these concerns are shared more widely. After it has been established that most of these concerns are held globally I ask if there is any process which is working on all these concerns. Agenda 21 is normally given as one of the answers, but if it isn't, I now bring it in to the discussion. If someone has mentioned Agenda 21 then I let them say what it is, and invite others to help if they know anything about it. I have found that there are usually at least two or three people who have some knowledge and often some experience of Agenda 21, and this is a good chance to let them show what they know.

I finish by showing the links between concerns raised and specific chapters in the Agenda 21 document.

There are a number of main points which I always make sure are covered, they are:

· Background: Over 150 governments signed the Agenda 21 document at the Earth Summit held in Rio De Janiero, 1992.
· Two-thirds of the changes outlined in Agenda 21 need to happen locally.
· Agenda 21 is as much about social justice as it is about environmental issues.
· It signals that new opportunities exist – governments are clearly saying they can't solve the problems by themselves.
· It provides a 'common language' which can build bridges between groups that would not usually interact.
· Local communities/projects can now approach councils saying how they can help the council to meet Agenda 21 obligations. So how will the council help them? This is different from the usual 'cap in hand' approach.
· Agenda 21 acts at different levels – global, European, national, local government, neighbourhood, personal.
· I cover most of the chapters to give students an idea of the range. This usually comes when making links between their concerns and Agenda 21.

○ *Further activities*

· If most of the participants are from the same locality, then I try to get the LA21 officer along towards the end of the course. This is an opportunity for students to find out what else is happening locally, and also creates a useful link which may help them after the course.
· '21 for 21', this is a phrase I picked up from an LA21 officer in Calderdale. It is where people think of 21 changes they will make to their lifestyles early in the 21st century. Although I haven't tried it yet, I will be incorporating this into my next course as a piece of home research towards the end of the course. It would also fit in well to the 'Where Next?' session.

○ *Bibliography*

1 Rescue Mission Planet Earth: A Children's Guide to Agenda 21, Peace Child International, 1994
2 Earth Summit '92, Quarrie, J (Ed), Regency Press Corporation, 1992

Beginnings and endings

Beginnings and Endings

This section includes the sessions used at the beginning and end of a course, and includes sessions such as 'A Tale of Two Chickens', which is often used to provide an early clear example of permaculture.

For a more detailed look at how to prepare for the first day, see
· 'TheFirst Day', pages 12-19
· 'Convenor's Guide', Pages 348-360.

⇨ Introductory Session

Cathy Whitefield

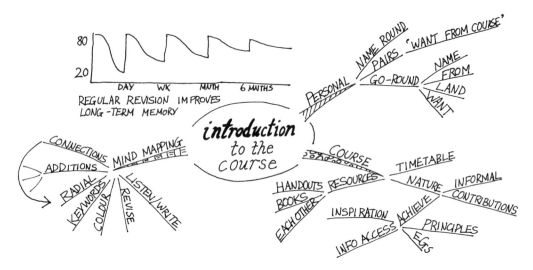

○ Objective

To give all course participants time to introduce themselves. To explain the nature of the course, housekeeping issues and the timetable. To introduce mind-mapping as a strategy for learning and note taking.

○ Learning outcomes

By the end of this session students will be able to:
- · Know why they and others are attending the course.
- · Help with housekeeping and domestic arrangements.
- · Explain what a mind-map is, and start to make their own.
- · Begin the course!

○ Context

The first session of the course.

○ Duration

60 minutes

○ How I teach this session

If there is more than one teacher on the course they should all be present for this session, and different ones can lead different parts of it. There are three parts:
- · Personal introductions.
- · Introduction to the course.
- · Mind-mapping.

Personal introductions

20 minutes

· Welcome.
· Name round. Each person says their name and where they live.
· Pairs. Students pair up with whoever is sitting next to them and spend 10 minutes discussing:
   a) anything from my daily life which is in my mind and which I need to drop in order
      to be here fully;
   b) what I want from this course.
· Go-round. Each person in turn:
   a) repeats their name and where they live
   b) says whether they have access to any land, eg garden, smallholding, farm
   c) says what they want to get from the course.

There's no time limit for the go-round, but in practice people are brief. The teacher or teachers should include themselves in it. The convenor and any other staff, such as cooks, usually introduce themselves more briefly.

Learning Names

Many people, including me, find name games tedious and embarrassing. Students get to learn each other's names soon enough on a Design Course, and on a weekend course they don't usually bother. But it's part of the teacher's role to know each person's name from the start. I have evolved a way of learning names which other teachers may like to try.

As each person says their name I look at their face and repeat it silently to myself. I do this both during the initial name round and the go-round later on. During the latter I have the time to repeat their name several times. I'm able to do this without missing anything they say. I also write their name down, together with notes of their replies to the questions.

If I get a chance during the go-round, I quickly look round the faces I have already learned, repeating the name with each face. This has to be done with care, so as not to look away from anyone who is speaking. I do it again at the end of the go-round.

I find this method is at least 95% effective: at the end of the Personal Introductions, if there are 20 students I'll know the names of 19, if not all of them. Just to make sure, I make a point of repeating each person's name while looking at them a couple of times over the next 24 hours.
P.W.

An additional method is to give each student a sticky address label, and use them as name badges!
A.G.

Introduction to the course

20 minutes

· Housekeeping. The convenor runs through anything students need to know about the administrative and domestic arrangements.
· Timetable. Give a brief introduction to the structure of the course, and hand out timetables.
· The nature of the course. Give an outline of:
   · What we aim to achieve on the course.
   · It's informal nature.
   · How every person's contribution is valued.
· Resources. Mention: books for reference, books for sale, general handouts, optional handouts.

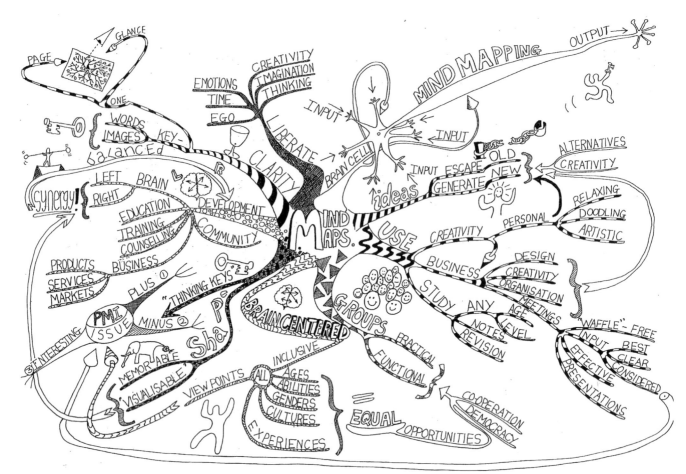

Created by Andrew Goldring and Wolf White for a Human Scale Development Initiative

Mind-mapping

20 minutes

This is introduced at this stage so that students can use it throughout the course. See mind-map! (and Refs 1 & 2.)

Bibliography

1 Use Your Head, Buzan, T, BBC, 1974*
2 The Mind Map Book

See also the 'Convenor's Guide', page 347-360.

Introductions/Teaching & Learning

Graham Bell

O Objective

To start the course well, co-operatively and safely.

O Learning outcomes

By the end of this session students will:
· Know each other's names.
· Feel encouraged to participate fully in the course.
· Feel comfortable with sharing their aspirations, and respect those of others.
· Understand domestic arrangements, and health and safety requirements for the duration of the course

O Context

First day (see 'The First Day', page 12-19). 'The Flywheel Effect' usually follows it (see page 176).

O Duration

35 minutes

O How I teach this session

Discussion and giving notices
· Introductions – 15 minutes
· Names (use a good game) – 15 minutes
· State own qualifications to teach – 5 minutes

Points to note:
· Listening is the hardest and most important skill in developing ourselves and the projects we work with. We can never have too much practice.
· Each voice is valuable. If we hear each other's views well and respect them, then others will respect ours. We choose to respect others' limits.
· 'Right' and 'Wrong' are not necessarily useful labels. We can move the process along better by harvesting what is useful to us, and leaving aside things which are of little use to us. Debate rarely changes opinions, and consumes much energy. Notwithstanding the above, in the final analysis the teacher's decision is final!
· We are all contributors in the learning process, because we all have information to share. The teacher's role is to facilitate self-education.
· Support teachers. We can change conventional pedagogic patterns of hypothesis and dispute to favour mutual respect and joyful discovery.
· Don't be afraid to ask for what you need.
· If you can see something needs doing, give yourself permission to do it.

See also the 'Convenor's Guide', pages 348-360.

The Flywheel Effect

Graham Bell

○ Objective

To discuss how different approaches to learning offer different advantages, and give students an opportunity to state their own needs and expectations.

○ Learning outcomes

By the end of the session students will be able to:

· Name a range of approaches to learning, and decide which to use for which purpose.
· State their expectations clearly.
· Co-operate with the needs of the group.

○ Context

First day (see 'The First Day', page 12-19).

○ Duration

15-25 minutes

○ How I teach this session

Discussion after an initial short talk

"The efficiency of an engine can be increased by adding a flywheel. This is a rotating wheel of high mass, which uses energy to start in motion, but when in motion provides momentum which smooths out any peaks or troughs in the engine's performance. By the participation of a critical mass of people (everyone) in the process of learning, we can optimise the power of the teacher to achieve results. To do this well, all the components of the engine (us) must fit together well."

Answer the following questions:

· What did you expect to happen on this course? (5 minutes)
· What can you contribute to this course? (5 minutes)
· What do you want from this course? (5 minutes)

Exercise

Think about your needs and your surpluses. Which of these would you like to share with others in the group? List at least three in each column. Try to keep each side even! Don't be afraid to call on others, offers and requests: a process by which we all get richer. (5 minutes) End with feedback on this exercise.

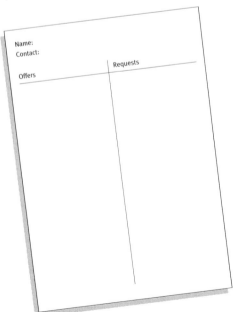

What Does it Mean to Be Alive?

Graham Bell

Why I include this session

It is vital to practise observation, however there is a subtle debate raised by this session. How do we know if something is alive? This is an important question because permaculture presupposes that biological processes are the prime focus of designers.

Objective

During this session students experience the value of observation, and question what it means to be alive.

Learning outcomes

By the end of this session students will be able to describe important aspects of what it means to be alive.

Context

Can be used on the first day (see 'The First Day', page 12-19). This is a good exercise to use in the graveyard shift after lunch, or placed before a break, with feedback afterwards.

Duration

45 minutes

How I teach this session

Walking observation

30 minutes

"In groups of four, walk for half an hour and look at what is around you. Discuss what it means to be alive and how the evidence of your senses suggests the answer for you. As a group be prepared to offer two points for a discussion at the end."

Allowing 30 minutes is usually enough. The teacher has a chance to join in, or to review the planned timetable against any feedback received so far today.

Discussion

15 minutes

The returning folks have a relaxed opportunity to swap views. We are now focusing in on WHAT REALLY MATTERS in life. This is a key link into concepts of sustainability, and is going to take us later into basic needs, and in effect WHAT ARE WE DESIGNING FOR?

Benefits of this exercise include:

· Attention span maintained.
· Brains re-oxygenated.
· Opportunity to chat to other course students.
· Early exercise in observation.
· Early opportunity to articulate views in front of whole group.
· It's a BIG QUESTION and challenges fundamental beliefs to emerge.

Permaculture: A Personal Story

Graham Bell

○ Why I include this session

So that students understand the potential of permaculture to offer positive solutions in their own and other people's lives. To get away from the idea of there being an 'objective perfection'. There is no permaculture rule book. I give my own personal interpretation, and students are encouraged to develop theirs.

○ Objective

To show visually, evidence of learning opportunities, positive practice and mistakes and what they teach us.

○ Learning outcomes

Students will have real life examples of permaculture and know how to find some for themselves.

○ Context

First day (see 'The First Day', page 12-19). Can also be used on the evening before the course starts .

○ Duration

Not more than 60 minutes.

○ How I teach this session

Illustrated talk

Images relate to personal history, discovery, observation, and working examples.
Points to note include:

· We all arrive at any new discipline with all of our life's experiences intact. Therefore none of us are total novices, and we all have experience and insight to share.
· Just as none of us are ignorant, none of us know it all, so we never stop learning.
· Observation is a good starting point in learning design skills.
· Everything is an opportunity.
· Permaculture is about taking personal responsibility.
· Permaculture is different in each realisation, because it is specific to time and place, and therefore it can only ever be a story particular to the people and the setting in which it is seen.
· Aspects of permaculture can be seen all around us.
· There is no such thing as a complete permaculture.
· Permaculture is a direction not a destination. Is a system moving in that direction? If yes, then we're building a permaculture approach.
· It's permissible to harvest ideas from anywhere, anyone, anytime. Good ideas are for replication.

⇨ What is Permaculture?

Graham Bell

○ Objective

To create a definition of permaculture from the existing knowledge of the class.

○ Learning outcomes

By the end of the session students will be able to:
· Name the main strands which distinguish permaculture design.
· State why this approach is needed.

○ Context

First day (see 'The First Day', page 12-19).

○ Duration

45 minutes

How I teach this session

Instructional talk

45 minutes

I present a number of points about the state of the world and invite contributions. I then ask students to get into pairs and give six points which answer the question: What is permaculture?

After this has been done, students are invited, in turn, to give a point each.

They have a sheet to record the points raised, thus:

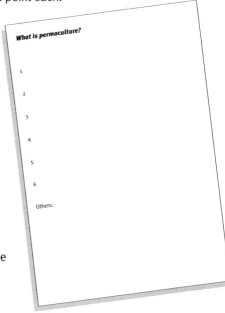

This exercise above is followed some days later, with a shorter exercise. I give out the following handout, which I ask them to complete. This takes five minutes to do, and five to review.

Here are some definitions:
· Permaculture is an ethical method for designing systems to meet human needs which are accessible to all.
· Permaculture is a patterned approach to problem solving which is transportable between cultures, locations and times.
· Users of permaculture strive to learn from and emulate the example of natural systems.
· Permaculture designs centre on connecting energy flows to maximise the harvesting of useful processes. All the material products we need also derive from such processes.

- Good permaculture designs eliminate waste by reconnecting unused products back into the system through careful planning, so they are assets rather than liabilities.
- Permaculture practitioners seek to maximise productivity whilst minimising input needs (work, money, people time, energy).
- Good design is achieved through appropriate placement and choice of elements in a system.
- Successful designs start small and start locally ('own doorstep outwards').
- Permaculture designs are evolutionary rather than revolutionary, favouring starting with what we have and progressing through to desired solutions.

Now complete the following:

My one sentence definition of permaculture:

_____

_____

_____

Why Do We Need Permaculture?

Graham Bell

○ Why I include this session

This session uncovers prejudices and strong opinions. It creates discussion and gives me a chance to spot early on who will dominate them, so that I can use friendly strategies to give others more talking space. It also uncovers people's distress about the world and enables me to spot who may need more support later on.

○ Objective

To put permaculture into a global perspective.

○ Learning outcomes

By the end of the session students will be aware of some of the complexities around moral and environmental issues as well as their own point of view.

○ Context

Can be used on the first day (see 'The First Day', page 12-19).

○ Duration

45 minutes

○ How I teach this session

Individual exercise

25 minutes

I provide a set of paired statements as a handout -see next page, and say "On your own, compare the following pairs of statements and decide which is nearest to your own view."

○ Discussion

20 minutes

After everyone has done this, discussion follows...

The instructor's role is to monitor and move the debate around, not to lead it. It offers you a good opportunity to gauge your class, and to deal with conflicts of opinion early and constructively.

1 ☐ Demands on the planetary ecosystem are outstripping supply
   ☐ There is enough for all if shared fairly

2 ☐ World-wide, traditional social structures and values are eroding
   ☐ New positive paradigms are needed for changing times

3 ☐ Many factors restrict our actions
   ☐ Terms and attitudes which convey those restrictions, eg predator, problem etc, convey a particular
     viewpoint, which can be changed

4 ☐ Modern agriculture can cope with increased demand
   ☐ World population is rising, whilst the area of good quality agricultural land is shrinking

5 ☐ Conflict is the inevitable result of inadequate resources
   ☐ Permaculture can supply all our basic needs

6 ☐ Competition is a fact of nature
   ☐ Co-operation is much better than competition

7 ☐ We need good leadership
   ☐ There are too many people in the world telling us what to do

8 ☐ Science can overcome problems that seemed insurmountable a few years ago
   ☐ Don't put too much faith in scientists

9 ☐ I have little control over the world at large
   ☐ The media love to exaggerate world problems

10 ☐ Worrying never changed anything
   ☐ We over-rate the importance of humanity in the world

This is a tick box exercise (see Toolbox, page XXX). The list should be no less than six nor more than 10 statements.

A Tale of Two Chickens

Patrick Whitefield

○ **Why I include this session**

Permaculture is notoriously difficult to define. The best way to get across what it's all about is often an example, and there is no better example than chickens, mainly because they can have so many beneficial relationships with other elements. The session also gives some useful information on keeping chickens.

○ **Objective**

To help students understand what we mean by permaculture.

○ **Learning outcomes**

By the end of the session students will be able to:
· Understand the meaning of permaculture.
· Approach the Principles session with a feeling of familiarity – see 'Links to principles and ethics', below.

○ **Context**

The very first session after the introductions, both on Design and Introductory Courses. The Principles session comes soon after.

○ **Duration**

45 minutes

○ **How I teach this session**

Introduction

2 minutes

I start by pointing out that this is just an example. You don't have to keep chickens in order to practice permaculture; vegan permaculture is quite possible.

Next I say we're going to look at how chickens are kept under both a battery system and a permaculture system. Specifically we're going to look at the needs/inputs and yields/outputs of chickens and see how they are provided and used in both these systems.

Brainstorm

10 minutes

INPUTS

E FOOD ↓
E SHELTER ↓
E WATER ↓
  GRIT
X OTHER CHICKENS ✓
E FRESH AIR ✓
? HEALTH CARE ✓
E PROTECTION ↓

OUTPUTS

✓ EGGS ✓
✓ (MEAT) ✓
P MANURE ✓
P FEATHERS ✓
P NOISE ?
P CO_2 ✓
P WARMTH ✓
X CULTIVATION ✓
X PEST CONTROL ✓

I draw a chicken top centre of the board and write Inputs to the left of it and Outputs to the right. First I ask the students to suggest what are the needs of the chicken, and I write what they say in a column beneath Inputs. Then I do the same with the outputs, reminding them that outputs we don't consider beneficial can be included.

Battery systems

5 minutes

I go through both lists, outlining how each input and output is treated in a battery system. On the inputs column, I write an E beside every need which is met by the expenditure of a great deal of energy. On the outputs column I put a P beside every output which becomes a pollutant. I draw attention to the fact that most entries on both lists get either an E or a P. Needs which are not met and outputs which are simply not used get an X.

Permaculture systems

20 minutes

Before turning to the permaculture chicken I point out that there's actually no such thing. Permaculture is not a definable standard like organics, with its symbol, but a way of thinking. None of the ideas I am about to mention is mandatory in permaculture, and individuals may use as many or as few of them as fit their own situation.

I draw the permaculture chicken's environment, starting in profile with the chicken house. To this I add:

· water butt
· attached greenhouse
· forage system.

I give the criteria for selecting chicken forage trees and shrubs, and a few examples. Legumes come into it, and with them the concept of multiple function.

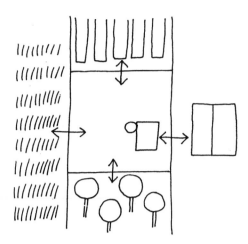

WORK = AN INPUT NOT MET BY THE SYSTEM
POLLUTION = AN OUTPUT NOT USED BY THE SYSTEM

Next I draw the chicken forage area in plan view, and one by one draw in the other elements of a small-holding, describing their relationship to the chickens as I go:

· house – frequent visits
· vegetable garden – tractoring
· orchard – pest control
· wheat field – gleaning.

I emphasise the importance of spatial design in making these relationships efficient or even possible.

Conclusion

10 minutes

I return to the two lists. First I show how permaculture has made use of many more outputs than the battery system, adding in any which did not get mentioned in the brainstorm but which have come up in my description of permaculture chicken keeping, eg CO_2, cultivation, pest control.

Then I show how permaculture, by linking up unmet needs with unused outputs, has reduced the energy input to the system and all but eliminated the output of pollution. I quote Bill Mollison's two equations (ref 1):

Work = an input not met by the system
Pollution = an output not used by the system

I point out that a common belief is that there are only two ways of producing food, or any other product: drudgery or the excessive use of fossil fuels. But permaculture offers a third way: design.

O Links to ethics and principles

· Earth Care, especially animal rights.
· People Care, a) the ethical implications of purchased chicken food, b) sustainable food production need not involve drudgery.
· Limits, the seasonal nature of sustainable production and consumption.
· Perennials, the forage trees and shrubs.
· Relative location, a) between chickens and their food plants, eliminating human labour, b) between all the elements in the design, especially the chicken house and greenhouse.

- Zoning, between house and chicken house.
- Network, between all the elements apart from the house.
- Elevation, raising the water butt on a stand so it can be used by hose.
- Multiple sources, noted with regard to the water supply.
- Small scale: most of the ideas depicted will only work on a small scale.
- Input-output, linking.
- Multiple output, a) accepting all the outputs of an existing element, chickens, rather than only one, b) choosing elements partly because they have a multiple output, eg legumes for chicken forage.
- Energy flows, evident throughout.

Mind-map

I write and draw mind-maps on the board as the session progresses. Obviously the lists of inputs and outputs vary somewhat from group to group.

Bibliography

1 Permaculture: A Designers' Manual, p24-25

Further research

Permaculture in a Nutshell, Chapter 2

Networking and Where Next?

Andrew Goldring

○ Objective

To provide students with an overview of how and where to get support, and ensure that they have a range of achievable goals and activities which they can begin when they get home.

○ Learning outcomes

By the end of this session, students will be able to:
· Find support appropriate to their needs.
· Register with the Academy.
· List goals and activities they want to pursue after the course.

○ Context

Last day. To be followed by course evaluation and presentation of certificates.

I run a 30 minute course review session immediately before this session, where a long roll of paper is put on the floor, with markings to represent each day and evening class. Students write and draw on the paper all the things they can remember doing during the course. I keep asking "What is missing?" and by the end there is a fairly complete record of the course in front of everyone. This helps to remind people of the range of opportunities, improves memory and provides a powerful way for me to see which exercises and activities have been most memorable. I can use this knowledge when designing the next course.

○ Duration

60 minutes

○ How I teach this session

Networking

20 minutes

The way I teach a course means that by this stage, students have built up a rich picture of the networks and activities in their area. So I start with a quick review and reminder. I do this as a game.

Game ...

Everyone stands in a circle with a ball. As someone throws the ball, they ask a question such as "Where would I find out about getting volunteers?", the person who catches the ball answers as best they can, and then asks another question and throws the ball to someone else. This continues for as long as seems useful...

... or Brainstorm

If people aren't aware of the range of useful networks, then you could do a brainstorm. The point is to make sure that people know where they can get all kinds of different help for any projects or initiatives, and who their allies might be. A useful way to stimulate the thinking is to ask what might be needed to achieve particular projects that they have in mind. (I find out if people have any projects they want to pursue, right at the beginning of the course.)

Short talk

I describe what support the Permaculture Association can offer. I stress that joining the Association is a simple and effective way to link up with other people who are also trying to design for sustainability. (See pages 362-365 for 'About the Permaculture Association'.)

Handout

I give out a sheet with useful permaculture addresses, such as sources of plant material, some of the projects that can be visited, sources of more specialist advice and any local contacts, groups or networks we may have built up.

Discussion

I ask what people know about mentoring, and whether this may be useful. I encourage each person to find someone who they can treat to lunch at least once a month in return for listening to them about their progress as designers and world changers. If the course is run locally, they can offer to do this for each other.

I also present pattern 147, Communal Eating, from The Pattern Language (Ref 1). This always sparks an animated discussion, as students see how all sorts of different possibilities can be woven into this pattern for informal networking.

Short break

'Where Next?'
40 minutes

Short talk, questions and answers

I start by drawing up the Action Learning cycle and give a short talk about the Diploma and how to enrol with the Academy. A handout about the Academy is made available (see page 366-367). I then answer any questions before moving on.

Goal-setting

Now we are at the goal-setting stage. I ask students to get out their charts from the 'Me, My Greatest Asset' session (see pages 304-306) and the Personal Health Plan that they created for homework. I ask them to spend a few minutes reviewing them.

Exercise – a letter to yourself
At least 20 minutes

There are all sorts of ways to structure goal-setting. One I have found useful is to ask students to think of setting goals in three areas: personal, social and environmental.

Setting out the instructions. "Set yourself goals in these three areas. They should be easily achievable, so that you can quickly build up learning and confidence which can then be applied to bigger projects later. Write them down in a 'letter to yourself'. At the end hand them to me, folded up, and we will review them at the follow-up weekend. Make a note of each goal in your notes so that you remember what they were!". If no follow-up session is planned, the letters can be sent to students after an agreed period.

Closing

I end with a round and invite students to share something they want to achieve with the rest of the group. Collaboration between students is encouraged.

Further activities

I set up a review weekend a few months later, so that goals can be reviewed against achievements, problems encountered etc. It is also an opportunity to refocus, reinvigorate and share post-course experiences.

Bibliography

1 A Pattern Language.

Further research

What Colour Is Your Parachute?, Bolles, R N, Ten Speed Press, updated annually. Invaluable guide to job-hunting, skills identification, and working out 'where do I want to go next?'.

Handbook for the Positive Revolution, De Bono, E, Penguin, 1992. Covers a wide range of issues vital to creating positive change, including self-help, making small but steady contributions, being effective and so on.

Aspects of permaculture design and practice

Aspects of permaculture design and practice

Some of the sessions in this section would be taught on nearly every design course, others may be used less frequently. It is a menu from which teachers can choose, depending on their own skills and experience, their students' interests, and the range of teachers or guest speakers available to deliver the course. These sessions will often be used to strengthen the understanding and provide clear examples of the central themes presented in Section 2.

The sessions within this section have been put into five chapters:

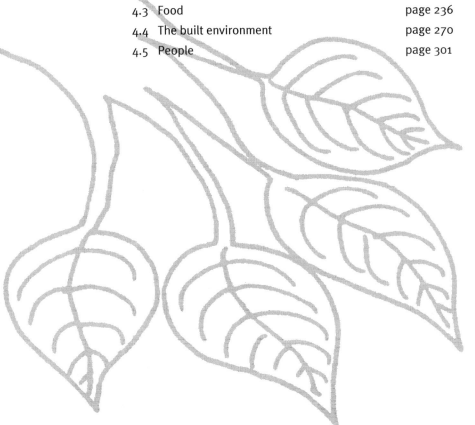

The Natural Environment

Natural Systems as Model and Inspiration
Micah Duckworth

Why I include this session

Through the observation of nature by Bill Mollison, David Holmgren and others, notably Eugene Odum, in the 1960s and 1970s, permaculture has grown out of a number of significant studies on ecology (Refs 1, & 2).

These studies, especially as a science, have provided a wealth of material on types of ecology, drawn from the patterns and processes within natural systems. The development of the application of this knowledge as models for designed solutions, including works by Paul Hawken, Michael Hough and Ian McHarg, has provided a solid basis for the practice of permaculture design (Refs 3 & 4).

This session provides an opportunity to explore these points and give the background for permaculture.

Objective

To gain an understanding of the features, functions and relationships within natural systems and how these inform our designed approach to sustainable living through permaculture.

Learning outcomes

By the end of the session students will be able to:
· See the origins and growth of permaculture within the wider historical context of academic study on natural systems and ecology.
· Explain why nature is our greatest teacher and model for sustainable design.
· Have a greater understanding of natural systems, what characterises them, and the different environments and levels in which they operate.
· List some of the principle features of natural systems that directly inform permaculture design, giving rise to zoning, sector analysis, stacking, edge etc.
· See how the permaculture ethics can be understood in terms of inter-related personal, social and environmental ecologies.
· Explain how the 'Mollisonisms' follow from an understanding of the ethics in this context.

Context

This should be taught near the start of a course, session 2 or 3, as it is so closely connected to permaculture principles. The session helps to enhance their value and give them a historical context through the origins of permaculture. The subject of natural systems may well be followed by looking at pattern, its language and understanding. I make many references to this session throughout the course. A walk in woodlands directly before this session is very helpful. Indeed most of the points made below could be covered during an outside observation exercise, with a short follow up session in the classroom.

○ *Duration*

120 minutes including a 15 minute break.

○ *How I teach this session*

Introductory round and 'ice-breaker'

5 minutes

I begin by perhaps asking each person to tell the rest of the group the last time they had a close encounter with nature.

Brainstorm – whole group

20 minutes

What characterises a natural system? ie inter-connectedness, diversity of elements, long-term stability, succession, evolution, adaptive, productive, cycling, edge, balance, energy efficiency, niches, stacking. I emphasise these points as features of good sustainable design.

Interactive talk

35 minutes

To answer the question, "How do we set about using these features to create a sustainable world?", I introduce the dimension of time, and draw a diagram to show how nature tackles it (see Weeds to Trees diagram below).

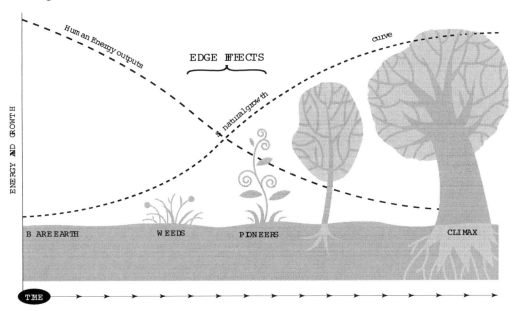

The weeds to trees model is important in describing the overall process of natural regeneration. Many important features are described:

· The classic natural S shaped growth curve.
· The idea that as nature takes over from bare earth to forest ecosystem (climax), required energy inputs decrease (inverse of growth curve).
· The value of edge and stacking effects.
· The role of people as pioneers in the greening of the planet. This introduces rolling permaculture, and empowers people by understanding that they can each take manageable steps towards great co-operative achievements.

Another diagram may be drawn showing the plan view of the same process. This introduces the concept of a natural polyculture and, with reference to the scale of energy transactions from ground to canopy, the origins of the zoning and sector models. (See next page).

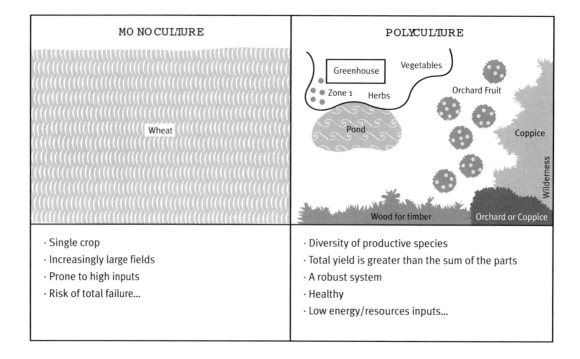

MONOCULTURE	POLYCULTURE
· Single crop	· Diversity of productive species
· Increasingly large fields	· Total yield is greater than the sum of the parts
· Prone to high inputs	· A robust system
· Risk of total failure...	· Healthy
	· Low energy/resources inputs...

An illustration such as the trophic pyramid diagram (Ref 5, p29) can be used to show how natural relationships give rise to larger structures such as webs, networks and cycles.

Break

15 minutes

Interactive talk

20 minutes

In the second half of the session, the group look again at the permaculture ethics, partly as a recap but also to introduce the idea of the ethics relating to a set of interrelated ecologies (see below):

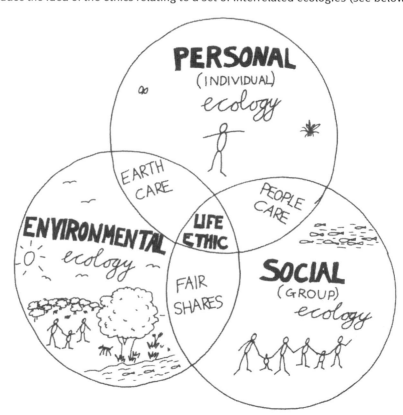

Personal: human mind, body and spirit, also includes individual plants and animals.

Social: human family and community, also includes plant and animal groups, eg guilds and herds.

Environmental: the whole thing! Forest, river, ocean, atmosphere, settlements etc.

I make the point that the ethics are primarily about care, responsibility and co-operation, focused on each of these areas.

Ecological health is ensured through keeping a balance within natural limits. Fair shares, or limits to consumption, are the means by which we manage this healthy balance of energy and resources between each form of ecology.

James Lovelock's Gaia theory, which sees this planet as a unified living organism, is a useful model to put forward when considering ideas like the health of the Earth. (See diagram below, Gaia model of 'spheres', and Ref 6.)

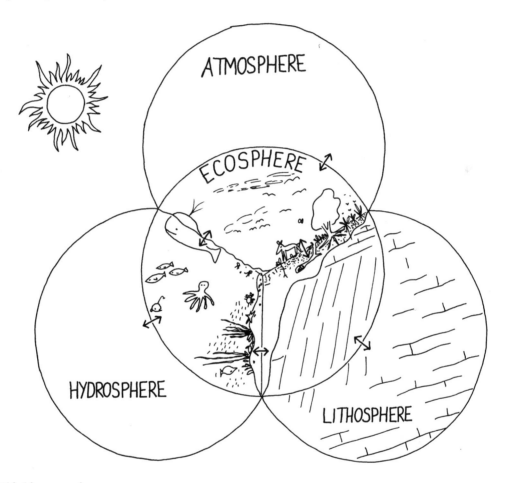

Thinking exercise

15 minutes

I ask the questions, "How does imbalance in one ecological level affect the others?" and "How is balance maintained in a healthy ecology?". I map all contributions on the board.

Making links

10 minutes

Links are made between the group's answers to reinforce the permaculture principles; particularly those which Mollison identified as:

· With nature, not against it.

· Everything gardens – everything has its time and place.

· Theoretically, there are no limits to yield.

· The problem is the solution.

· Apply minimum effort for maximum effect.
· Every element serves many functions; every function is supported by many elements.

This is a good exercise for linking ethics to principles, and lays the foundations for following sessions to pick up on more specialised subjects.

Living examples

Everywhere! Emphasise observation, and nature as teacher.

Also, examples of designs which have been directly modelled on, or make use of, natural systems: forest gardens, reed beds and 'living machines' (water treatment), Bon Fils and Fukuoka growing methods.

Further activities

Outdoor observation exercises (see 2.4 'observation' section) which contribute to an increased understanding of what features occur in natural systems and how they interrelate.

Outdoor activities can help to make the important point that in addition to its formative role in the development of permaculture, the study of natural systems should be understood as an on-going necessity, related to the need for inclusion of wilderness (Zone 5) in any site design or land use plan. Apart from its inherent value, land left to nature (wilderness) acts both as an indicator of environmental sustainability, and connects us directly to the living systems which support us.

Agenda 21 link

Permaculture's emphasis on natural systems, modelling and patterning, is fundamental to ensuring that people understand the basis behind Local Agenda 21; that is, to enhance the capacity of humans to live in harmony with nature.

Bibliography

1 Fundamentals of Ecology, Odum, E, W B Saunders, 1971
2 Permaculture One
3 Designing with Nature
4 The Ecology of Commerce
5 Permaculture: The Designers' Manual
6 Healing Gaia: Practical Medicine for the Planet, Lovelock, J, Harmony Books, 1991

Further research

The Web of Life
The Earth Manual – how to work wild land without taming it, Margolm, M, Heyday Books, 1985
A Sand County Almanac, Leopold, A, Sierra Balbutine, 1966
Cities and Natural Processes, Hough, M, Routledge, 1995
'Stability, Change & Harmony': A Permacultural View , Issue 6, Tyler, E, Permaculture Magazine, Permanent Publications, 1994

Biodiversity/Zone 5

Patrick Whitefield

In Australia, where permaculture originated, Zone 5 is defined as wilderness: land which has not been modified by human activity and which is left alone, both for the enjoyment of the wild plants and animals which live there, and as a source of knowledge and inspiration for us.

Here in Britain we have no such land. Even the Scottish Highlands have been drastically modified by human activity: they were once largely forested. Many of the most biodiverse habitats here are ones in which humans have played an active part for hundreds or thousands of years, such as unimproved hay meadows or coppiced woodlands. Leaving these areas to their own devices may actually lead to a decrease in biodiversity.

So what do we mean by Zone 5 in a British context? I feel the best working definition is: those areas of land where the interests of wild plants and animals have priority over those of humans. We may continue to play our part in the ecosystem of such areas or we may decide not to; but whatever we, do it is in the best interests of biodiversity.

I realise that this definition raises more questions than it answers. Indeed that debate is central to how I teach the subject.

Why I include this session

The purpose of permaculture is not to cover the Earth with 'edible ecosystems' designed to meet our needs. It's to so intensify our own production that we need less land to meet our own needs, thus releasing much of it for the enjoyment of all the other species we share the planet with and which have as much right to flourish here as we do.

Objective

To encourage thought and discussion on how we should approach the conservation of natural biodiversity in Britain, and give some basic information on the subject.

Learning outcomes

By the end of the session students should:
· Have had an opportunity to reflect on their own stance on wildlife conservation, and to compare it with that of others.

They will be able to:
· Explain the importance of Zone 5 considerations in permaculture design.
· Name some of the semi-natural habitats which may be encountered in Britain.
· Give an idea of the degree of wildlife habitat destruction which is taking place.
· State some of the important factors in wildlife assessment.
· Know where to get free expert help in assessment, and advice on working with wildlife.

○ *Context*

This session could be taught at any point in the course before the main design exercise. Since I work through the zones from inside to out, I end up doing Zone 5 just before the main design exercise.

○ *Duration*

90 minutes, but I don't mind over-running if it's arousing a lot of interest.

○ *How I teach this session*

Part 1: Discussion

30 minutes

I hand out photocopies of two articles with contrasting viewpoints:

· 'Wildlife Habitats & Permaculture'
· 'Letting Go of the Countryside' (Refs 1 & 2).

Everyone reads them, and I ask for comments. This usually leads to a lively discussion around the theme, 'What is Zone 5 in a completely human-influenced landscape like Britain's, and how do we work with it?'. I draw the discussion to a close according to how it's going rather than by the clock.

Part 2: Talk

30 minutes

The length and content of the talk vary according to: a) how much time the discussion took, and b) how much information was covered in it.

There are three main themes:

General points:

· Every design should have an element of Zone 5 in it.
· Why? Three main reasons for conserving biodiversity (see box).
· How? Three main ways of doing it.

Habitats in Britain:

· Mention main kinds of habitat and rate of habitat loss.
· Special mention of linear and urban habitats (Ref 3).

How to assess the biodiversity value of a piece of land:

· Where to get help.
· Six things to look for.

For detailed subheadings see the mind map below.

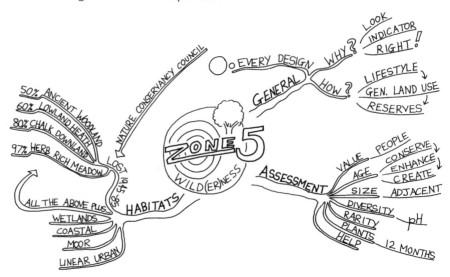

Why Conserve Biodiversity?

There are three levels on which we can answer this question.

Firstly, because we like to look at wildlife. Put like that it sounds trivial, but we do have a spiritual and emotional need at least occasionally to experience beautiful places, where wild plants and animals thrive and human material needs take a back seat.

Secondly, from practical self-interest. Wild species are indicators. If we create a world in which they thrive it's probably one in which we'll thrive too. If we try to create a planet on which nothing but us and a few species directly useful to us can survive, it's unlikely that we can survive either.

Also, the gene pool of wild plants and animals represents an incredible resource, including undiscovered medicines, new food crops, new germ plasm for existing crops and much else besides. We destroy them at our own peril.

Thirdly, because it's right. Wild plants and animals have just as much right to exist and thrive as we have. This right is not dependent on their usefulness to us.

Each of these reasons is valid, but I feel the third is the most fundamental. It is also the closest to my heart.

Part 3: Slides

30 minutes

Approx 24 slides, illustrating:

· A variety of habitats, including: pond, hedgerow, coastal, grassland, lawn, woodland, urban.
· Effects on biodiversity of: nutrients, mowing, coppicing, succession, age, exotic species.
· Contrast between rural and urban situations.
· Habitat 're-creation'.

○ Agenda 21 link

Chapters 15 & 17.

○ Bibliography

1 'Wildlife Habitats & Permaculture', Thomas, B, Permaculture Magazine, Issue 2, 1992
2 'Letting Go of the Countryside', Evans, M, Permaculture News, No 22, 1991
3 The Ecology of Urban Habitats

○ Further research

Future Nature, Adams, W M, Earthscan, 1996. An overview of nature conservation in Britain
Farming & Wildlife, Andrews, J, & Rebane, M, RSPB, 1994
How to Make a Wildlife Garden, Baines, C, Elm Tree Books, 1985
The Wild Flowers of Britain and Northern Europe
The History of the Countryside
Practical Conservation, Tait, J et al, Open University, 1988. Broadscale design methods for aesthetics and nature conservation

'Wilderness', Issue 7, Permaculture Magazine, Permanent Publications, 1994

BTCV Publications – a range of practical guides

British Trust for Conservation Volunteers (BTCV), 36 St Mary's St, Wallingford, Oxon OX10 0EU
Tel: (01491) 839766

The Wildlife Trusts, The Green, Witham Park, Lincoln LN5 7JR. Tel: (01522) 544400. Umbrella
organisation for the County Wildlife Trusts

English Nature, Northminster House, Peterborough PE1 1UA. Tel: (01733) 455000

Scottish Natural Heritage, 12 Hope Terrace, Edinburgh EH9 2AS. Tel: (0131) 447 4784

Countryside Commission for Wales, Plas Penrhos, Ffordd Penrhos, Bangor, Gwynedd
Tel: (01248) 385500

Irish Wildlife Federation, 132A, East Wall Rd, Dublin 3. Tel: (00 353) 1 366821.

RSPB, The Lodge, Sandy, Bedfordshire SG19 2DL. Tel: (01767) 680551.

Biodiversity

Mark Fisher

Why I include this session

Permaculture as an earth science arose from protracted observation of natural systems. Thus biodiversity, and the relationships which hold it all together, was the inspiration for permaculture, and the natural world continues to teach us how best to design self-sustaining systems. We are urged, as permaculturists, to create and conserve wilderness habitats and species refuges to maintain that biodiversity. We do this because we recognise that the quality of our future existence, and the range of options open to us, is tightly linked to maintaining the greatest diversity. In these senses, biodiversity is a key element of permaculture.

Objective

To recognise the importance of biodiversity for sustainability and how permaculture systems maintain it.

Learning outcomes

By the end of this session, students will be able to:
· Begin to read landscapes and understand why the habitats within them are distinctive.
· Start to appreciate the natural relationships between soil, micro-organisms, plants and animals and how these relationships can be harnessed in permaculture systems.
· Recognise the personal and collective responsibility for maintaining biodiversity and develop practical skills that contribute to this.

Context

The principles of natural systems is often an early topic in the Design Course and its purpose is to set the context of much that follows. Biodiversity may be addressed in that topic but it is worth considering returning to it at a later stage and potentially combining or linking it with bioregions. By bringing it later to the course, the opportunity is given to see more clearly how permaculturists can use their knowledge and skills to better maintain biodiversity.

Duration

120 minutes including at least one break

How I teach this session

Biodiversity is often taught with reference to global conventions, and national and local action plans. While the documents arising from these sometimes contain broad strategies, they very quickly focus down to endangered species and habitats and in the process lose the holistic emphasis that is key to permaculture.

A more rewarding approach is to examine and compare a range of natural habitats and develop a reading of the elements in those landscapes. These can be called Lessons from the Natural World and it is a process of discovery and identification. The rationale is that the student begins to appreciate diversity and is able to recreate or extend those habitats. It will be remembered that many students only have access to urban landscapes for this and it must be shown that conserving and creating wildlife habitats and species refuges are as important there as anywhere else.

This approach is best delivered with a slide show and is complimented by practical work and handouts. Thus this session requires considerable resources which may present a challenge to the teacher.

Slide show

A representative slide show would include the following habitats:

- Upland acid moors
- Limestone upland
- Coastal cliffs and upland
- Woodland
- Waterlands
- Grassland (pasture and meadow)
- Seashore
- Lowland heath

Within those habitats, I show plants and animals that are indicators of soil type and local climate. If there are any known properties and uses such as wild foods, dynamic accumulators, coppicing, suitability for hedging, then these are pointed out. Symbiotic relationships and nutrient recycling can be covered using fungi, particularly in relation to woodlands, and nitrogen fixation in relation to poor soil and seaside dunes. The principles of permaculture can be confirmed, such as stacking, as shown by a natural forest edge.

I conclude the slide show with illustrations of a locally-constructed wildlife garden with an appreciation of the habitats created, eg permanent water, pasture, hedgerows, stone walls, managed and wild woodland.

The slide show is complemented after the break with a classroom practical or design exercise.

Classroom practical

The ability to propagate plant material allows people to provide their own resources for building habitats and refuges. Seed sowing is a necessary skill for all permaculturists but propagation through cuttings and divisions greatly increases the range of material that can be produced. Suitable classroom practicals to complement the slide show depend on the season but can include hardwood cuttings of trees and shrubs, and the division of useful perennials such as comfrey (Ref 1).

Design exercise

1 Brainstorm useful elements for inclusion in a wildlife garden
2 How can we assemble them in a way that maximises their potential to draw in wildlife?

Agricultural diversity can be increased by a range of strategies including:

· Landscape patterning: micro-climate and niches for diversity in space; regional development of ecosystems.
· Agroforestry: diversity in vertical use of resources – light, water, soil, nutrients etc.
· Multiple cropping: diversity as above; also labour spread out; provides stability of yield both economically and environmentally.
· Versatile rotations: diversity in time; diversity in demand on soil nutrient and water; soil improvement.
· Integration of animals: increase of diversity and complexity, more synergies possible.
· On farm seed selection: different varieties of one crop – ensure sufficient yield under varying climactic conditions; preserve gene base – open pollinated not cloning/F1 hybrid.
· Use of farm/field boundaries: Fences and field bunds can be used for alley cropping – useful plants which are different from the main field crops.
· Beneficial connections: companion planting; mixed cropping; relay cropping etc.

Mike Feingold

Links to ethics and principles

Many of the principles are exemplified in the topic including working with nature, everything gardens, stacking, maximising diversity and relationships.

Agenda 21 link

Chapters 15 & 17.
Local Biodiversity Action Plans are seen as the way of translating the national strategy (UK Action Plan on Biodiversity, 1994) into action. In many areas, these are part of a Local Agenda 21 process and there is guidance available on drawing up a local plan from the Local Government Management Board (now IDA) and the UK Biodiversity Group (Refs 2, 3, & 4).

Bibliography

1 Creative Propagation, Thompson, P, Helm, 1989
2 Biodiversity: The UK Action Plan, CM 2429, HMSO, 1994
3 Local Government Management Board (now IDA), Layden House, 76-86 Turnmill Street, London, EC1M 5QU
4 UK Biodiversity Secretariat, Dept. of Environment, Transport and the Regions, Tollgate House, Houlton Street, Bristol, BS2 9JD

Further research

Food for Free
Wild Flower Gardening, Stevens, J, Dorling Kindersley, 1987
The Small Ecological Garden, Stickland, S, HDRA/Search Press, 1996
The Natural Garden Book, Harper, P, Gaia Books, 1994
The Complete Manual Of Organic Gardening,Ed. Caplan, B, Headline Books, 1994

Regeneration

Chris Dixon

Why I include this session

Natural regeneration provides us with powerful models for our design work. The patterns unfolding in regenerating systems have applications not only in environmental design but also community and personal development.

To make effective use of the principle Work With Nature Not Against It requires us as designers to have an understanding of how nature works and what nature is trying to achieve in any given situation. Most of Britain was forested in the past (estimates range from 80-97% forest), and that's what most land is trying to turn back into today. The only reason land doesn't turn back into forest is due to the limiting factors that apply on any given site. Initiating regeneration is simple and generally requires only the removal of those limiting factors. The initiation of regeneration, and the observation and interaction with this dynamic process is a hugely empowering experience for participants.

Objective

To foster an understanding that natural regeneration processes can be understood and worked with.

Learning outcomes

By the end of this session students will be able to:
· Explain that land is dynamic not static.
· Explain the principle Work With Nature Not Against It.
· Recognise that the general pattern of natural regeneration in Britain is towards forest.
· Identify the three main factors in sustainable systems (water, soil, diversity).
· Explain the concept of steering.

Context

Regeneration can be taught at various stages in a design course. I have at times gone straight from ethics, Earth Care, or from principles, Work With Nature Not Against It, into regeneration.

Duration

60 minutes

How I teach this session

Outside observation
30 minutes
1 If I am teaching from my home I show students the regeneration project on site. I find this to be a powerful teaching resource. The site can largely tell its own story and it is only necessary for me to provide a commentary on what people are looking at, answer specific questions and point out examples or evidence for the main points.
2 If I am teaching in another locality I try to find examples of regeneration nearby. This is usually straightforward as nature is always trying to colonise everything. Possible sites include any so-called

derelict land, railway cuttings, building sites, abandoned quarries, overgrown gardens, the cracks between pavings, roadside verges. Once students' attention is focused upon the concept of regeneration they will find many examples.

Slide show

10 minutes

As well as site work, or if a site is not available, I use a set of slides to show some of the stages in the regeneration process.

Presentation

20 minutes

I also generally make some form of presentation of succession on a white/black board. I have used an Ice Age model (see diagram), which shows the movement from mineral rich but organically poor soils after the ice age through to high forest. I draw this as a curve and include quick sketches of plants and animals.

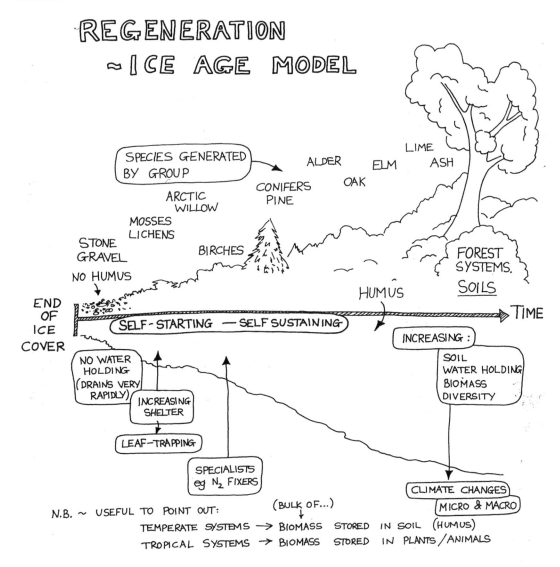

REGENERATION
~ ICE AGE MODEL

SPECIES GENERATED BY GROUP

ALDER   ELM   LIME   ASH
OAK

CONIFERS PINE

ARCTIC WILLOW

MOSSES LICHENS

STONE GRAVEL

BIRCHES

NO HUMUS

FOREST SYSTEMS. SOILS

HUMUS

END OF ICE COVER

SELF-STARTING — SELF SUSTAINING

TIME

INCREASING:
SOIL WATER HOLDING BIOMASS DIVERSITY

NO WATER HOLDING (DRAINS VERY RAPIDLY)

INCREASING SHELTER

LEAF-TRAPPING

SPECIALISTS eg N$_2$ FIXERS

CLIMATE CHANGES
MICRO & MACRO

N.B. ~ USEFUL TO POINT OUT:
(BULK OF...)
TEMPERATE SYSTEMS → BIOMASS STORED IN SOIL (HUMUS)
TROPICAL SYSTEMS → BIOMASS STORED IN PLANTS / ANIMALS

I FIND IT VERY USEFUL TO DRAW OUT THE TIMELINE AND THEN JUST START WITH "END OF ICE COVER - MINERAL RICH, WELL DRAINED GRAVELS", AND GET THE GROUP TO SUGGEST WHAT HAPPENS NEXT. WITH STEERING I USUALLY FIND IT EASY TO DRAW OUT ALL THE MAIN POINTS AND, AS USUAL, THE GROUP WILL COME UP WITH SOME INTERESTING STUFF I HADN'T THOUGHT OF

I usually get the students to drive this presentation by asking what comes next in the succession. There are no wrong answers here as I can fit any contributions within the basic curve and keep prompting until gaps are filled. Great detail on species and precise orders of succession are not crucial here. The presentation allows me the opportunity to make the main points if they do not arise from the group.

I point out that there is no single line of succession, that on any site the process will be complex and flexible, depending on the limiting factors relating to a specific site and how much intervention is made. Also, after a few years, examples of all stages of succession will be present on the same site at the same time.

I present a list of species and the order in which they occur, making reference to the process of clumping. Non-grasses form clumps around individual trees or shrubs, especially gorse. Over time the clumps expand and join together forming denser stands, the thicket stage. Clearings become apparent amongst the clumps. Tree crowns open out above bracken and gorse and begin to shade them out. Spaces and forest floor appear under the enlarging crowns.

Main points

There are a number of points to be made regarding natural regeneration, including:

· Land is a dynamic ongoing process.
· Sustainability. Attention needs to be pointed to three main areas: water, soil and biodiversity.
· Limits to yield. Identification of limiting factors and their ordered removal. For example:
  · Grazing animals and agriculture.
  · Lack of available seed sources.
  · Tenacious species.
  · Lack of light.
  · Lack of water.
· Steering; in practice, the regenerating system can be steered toward any of the zones.
· Succession.
· Maintenance as harvesting.
· Not always appropriate, eg species rich meadows, wet lands.

Other comments

I generally broaden regeneration out at some point. I remind people that we are dealing with a holistic system which includes both matter and consciousness. In the same way that environments and soils have become eroded and degraded, so too have we as individuals and communities suffered erosion and degradation. I present the idea that regeneration of individuals and communities can be approached in the same way as that of environments. That is, we can think of ourselves and our communities as having similar tremendous powers of regeneration, able to recover from enormous damage, heal ancient wounds, and re-create our natural diversity and abundance.

Link to principles

There are a number of concepts or topics in permaculture design which are particularly suited to the communication of the holistic perspective, that is concepts which permeate all the ethics and principles. I think regeneration is one of these key concepts. All the principles can be shown to good effect.

Living examples

Tir Penrhos Isaf, Hermon, Llanfachreth, Dolgellau, Gwynedd, LL40 2 LL. See the Permaculture Plot and 'Tir Penrhos Isaf: Permaculture Design in Snowdonia', Permaculture Magazine, Issue 2, Permanent Publications, 1992.

Further activities

· On guided walks I take some pieces of fencing wire that have been bent into the shape of the sole of a boot, about size 10. If I come across something unusual, like an orchid, I place the wire boot around it and make a comment. This marks delicate specimens so they don't get trodden on and makes people more aware of what they can crush if they tread carelessly. I also get them to toss the wire boot onto the ground and do a species count within the outline. This forms part of our sensitivity training.

· This session can be expanded to relate to regeneration of communities and ourselves as individuals.

Agenda 21 link

Chapters 11(B), 12(B) &14(E).

NB: I am indebted to Phil Corbett for first making clear to me the similarities between erosion of soils, individuals and communities, (1991)

 # *Soil*

Patrick Whitefield

○ ### *Why I include this session*

Soil is fundamental not just to permaculture, but to the very existence of life on the planet. Although the permaculture approach to soil is outlined in the principles, a basic understanding of how soil works is also necessary.

○ ### *Objective*

To cover the basics of soil science from a permaculture viewpoint. To give students an opportunity to get to know soil physically.

○ ### *Learning outcomes*

By the end of the session students will be able to:
· Explain and/or discuss the concepts of: soil fertility, soil texture, soil organic matter, tilling and no-till systems.
· Have had an experience of handling and examining soil.

○ ### *Context*

Theory: first thing in the morning, usually on day four, immediately followed by the sessions on farming.

Practical: first thing after lunch on the same day, or occasionally the day after.

○ ### *Duration*

Theory: 60 minutes
Practical: 90 minutes

○ ### *How I teach this session*

Part 1: Theory
60 minutes

Handouts
a) a general handout giving, on one side, Soil Texture Test by the Hand Method, and on the other, Soil Indicator Plants, list and guidance.
b) an eight-page booklet, giving more information than I have time for in the talk, for sale at 50p.

Talk
This is the longest talk I give in any course I teach. It can only be done first thing in the day, and I can only get away with it because of the enthusiasm and passion I have for the subject.

There are many aspects of the subject which could be covered on the Design Course. Given the constraints of time I have narrowed them down to the following four, which I feel are the most important. The first three are fundamental to a holistic understanding of soil, while the fourth is especially relevant to permaculture.

What is soil?

· Proportions of the four main constituents.

What is soil fertility?

· Brainstorm the eight principal factors of soil fertility.
· Draw up a table showing how these relate to different soil textures.

Organic matter

· How it affects the factors of soil fertility.
· Applied to the surface or incorporated: the need for composting.

Tilling

· Reasons for
· Reasons against
  See the mind-map for detailed subheadings.

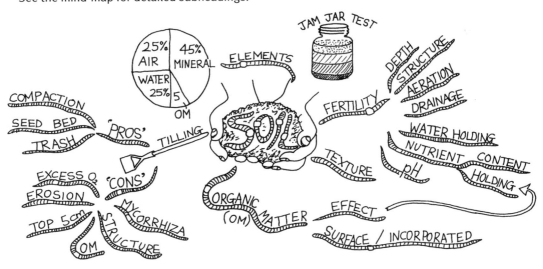

Slides

A slide set follows this session. The first few slides depict soil erosion, the rest illustrate permaculture approaches to farming.

Part 2: Practical

90 minutes

I have two main approaches to this practical, the indoor and the outdoor. The indoor is more appropriate in urban situations, and perhaps in winter. There is often time to combine the two.

The aim of the practical is not just to learn about soil, but to get people's hands in it, and start them making a relationship with it.

Indoor

Before the course starts I ask as many of the students as possible, usually via the convenor, to bring soil samples with them.

If possible I ask them to bring a number of contrasting samples, eg same soil type under different kinds of vegetation, or soils of contrasting textures. I ask them as far as possible to bring a complete profile including any litter or plant growth on the surface, but any handful of soil will do. I bring some samples myself so as to be sure of having something to work with.

We look at the soil profiles and deduce what we can from them, and do a texture test on each.

Outdoor

Ideally carried out on the main design site, but not necessarily. Either:

a) dig a number of soil inspection pits in places of contrasting vegetation, topography, drainage or soil texture, so as to get a variety of profiles. Examine the profiles and do texture tests. Or:

b) divide the students into small groups, get each group to dig one of the inspection pits, record what they observe, then have a look at each of the other pits in turn.

Agenda 21 link

Chapters 12(B) & 14 (E, J).

Further research

Soil Care and Management, Readman, J, HDRA/Search Press, 1991. Practical guide to soil for gardeners

Towards Holistic Agriculture, a Scientific Approach, Widdowson, RW, Pergamon Press, 1987. Chapter on soil gives a good basic outline of the subject

Organic Farming, Lampkin, N, Farming Press, 1990.Chapters 2, 3 and 4 give a fairly detailed overview of soil and crop nutrition from an organic perspective. Assumes some basic knowledge

Soil Erosion in Britain, Hodges, RD & Arden-Clarke, C, Soil Association, 1986

Soil & Soil Regeneration

Anne-Marie Mayer

Objective

To explain how soil works in natural and agricultural systems and the ways in which soils are affected by different practices. To explain how damaged soils can be regenerated.

Learning outcomes

By the end of this session students will be able to:
· Describe what soil is – its components and how it works.
· Explain how different practices and conditions affect soils and plant growth.
· List techniques that protect soils whilst maintaining productivity.
· List strategies to regenerate damaged soils.
· Make connections between soils and other natural systems.

Context

The topic is best presented after introductory sessions, principles and ethics but before detail on plants, trees and zones.

Duration

Between 120 minutes sessions (minimum) to a full day

If time is limited to 120 minutes I feel the essential components are as follows: introduction, spiral of destruction, natural soil fertility, good practice and soil regeneration. The other topics provide useful background and a clearer understanding. Students could be referred to text books for some of the more basic soil information. Practicals help illustrate the points and give more examples.

How I teach these sessions

I use a range of techniques, and where possible this includes the use of videos, games and practicals. I sometimes set homework observations eg different soil types around the area.

Introduction – interactive talk
30 minutes
To include context and problems worldwide, and how soils are produced.

Tribal soils classification – interactive talk
20 minutes
A wide range of facets including physical, biological, agricultural and environmental factors.

Definitions and descriptions of soil – interactive talk
30 minutes
Texture, structure, consistency and pH.

Soil components and interactions – questions & answers

20 minutes

Particles, water, air, organic matter and soil life, interdependence and relationship to structure.

Soil ecology, living soil – small group brainstorm followed by feedback

20 minutes

The range of soil life and roles eg bacteria and nutrient cycling, mycorrhiza and symbiosis with roots, earthworms and organic matter assimilation.

How natural soils work – interactive talk

20 minutes

Natural fertility, cycling of nutrients, ethylene cycle, where nutrients are held in the soil, how plant roots obtain nutrients.

Maintenance of soil fertility – interactive talk

20 minutes

Eg nitrogen cycle, different micro-organisms involved, conditions that cause N depletion, use of fertilisers, legumes.

Spiral of destruction – soils and modern agriculture – whole group brainstorm

20 minutes

Compaction, ploughing, loss of organic matter, soil erosion, artificial fertilisers, pesticides, loss of natural fertility, death of soil.

Good gardening/farming practice to maintain soils – small group discussion followed by feedback

30 minutes

Organic practice, keyline, use of perennials, minimum tillage, mulching etc.

Soil regeneration – interactive talk

40 minutes

Use of soil reconditioning unit, swales, remineralisation and connections to other topics – green manures, dynamic accumulators, trees, natural regeneration, arid landscapes.

Practical activities

I would always try to include a hands-on activity.

· Soil composition tests.

   30 minutes

   Finger test, glass jar test.

· Soil profile.

   30 minutes

   Dig a hole and time drainage at different sites.

· Observation

   20 minutes

   Observe plants growing in different sites and the link with soil.

· Web of life exercise.

   20 minutes

   Each person takes on the role of a soil minibeast or other soil element and stands in a circle. A ball of string is passed between participants to make the connections between the elements. Each person holds on to the string as it passes by and a complex web is formed.

The interconnectedness is illustrated when one person lets go and the web starts to collapse.

· Double dig practical and mulch exercise.

60 minutes

· Broadscale strategies.

40 minutes

Mark out the contour for swales using a water level.

○ *Link to ethics and principles*

· Cycling of nutrients in the soil. Cycle of life, death and decay and soil fertility.
· Integration/connectedness of soil components and soil life.
· Edge effects – the soil as an example of productive edge between earth and atmosphere.
· Succession – related to soil regeneration.

○ *Further research*

Soil Ecology, Killham, K, Cambridge University Press, 1994

Farming and Gardening for Health or Disease, Howard, A, Faber and Faber, 1945

The Nature and Properties of Soils, Brady, NC, Collier Macmillan, 1984

The Living Soil and the Haughley Experiment, Balfour, E, Faber, 1975

Ploughmans Folly, Faulkner , EH, Michael Joseph Ltd, 1945

Articles

'Soil Fertility', Harrison, L, Permaculture Activist No 26, 1992

'Historical Changes in the Mineral Content of Fruit and Vegetables', Mayer, AM, British Food Journal 99/6 pp207-211, 1997

Specific chapters from general books

Designing and Maintaining your Edible Landscape Naturally, Part 6, pp261-3

Permaculture: A Designers' Manual, pp182-226

Earth Users Guide to Permaculture, pp30-37

Videos

Global Gardener arid – desert strategies

Global Gardener temperate – Tasmanian soils and forests

All Muck and Magic, HDRA – episode 2

Micro-climate

Simon Pratt

○ Why I include this session

Some understanding of global and local climates, and of how we can modify the latter, are essential for permaculture design.

○ Objective

To explain how we can modify climate locally.

○ Learning outcomes

By the end of this session students will be able to:
·   Explain how climate is modified by local factors.
·   List strategies to modify local micro-climate.

○ Context

Early in the course, because it is important background to any design exercise. Useful links can be made between this session and to 'Buildings and Structures', 'Zone 1' and 'Windbreaks'.

○ Duration

40 minutes

○ How I teach this session

Presentation, questions and answers

40 minutes

I use a blackboard and/or OHP for graphical presentation. I find it preferable to speak from my own experience wherever possible. I ask students to give their own examples.

Key points:
·   Introduction:
    ·   Climate can vary between places close together, so it's vital to analyse site climate rather than rely on broad statistics for the area (Ref 3, p36).
·   Topography:
    ·   Slope and aspect.
    ·   Effect on solar radiation, shade, cold air drainage, thermal belt, winds (observe vegetation to determine direction) (Ref 3, p36).
·   Water:
    ·   Modifying temperature: heat store, evaporative cooling, eg fountains.
    ·   Reflecting light and warmth in winter (Ref 3, p41).
·   Structures:
    ·   Greenhouse, earth berms, walls, trellis, tyres and drums etc (Ref 3, p41).

- Soils:
  - Colour affects, absorption/reflection
  - Remove mulch in spring (Ref 3, p43).
- Vegetation:
  - Transpiration, convective transfer of heat.
  - Forest cool during day and warm at night.
  - Shade, sun trap, windbreaks (permeability, multiple functions), insulation (Ref 3, p43).
- Global climate:
  - Important to know that it exists, but beyond the scope of this course to cover in detail (Ref 1, Ch 5).
  - Sources of climatic data: temperature, precipitation, wind, day length (light) (Ref 2, p23).
  - Cutting rainforest and greenhouse gases can affect global climate; trees create rain.

Further activities

- Slides can be used to good effect and there are some micro-climate examples in the Global Gardener video (Ref 5).
- Students can identify and map micro-climates on a residential course venue site, or on their own sites for non-residential courses (see micro-climate study form in Ref 4).
- Observation of micro-climate on site can be included in experiential walks. This is especially good in extremes of weather – frosty day for frost pockets, cold winds for shelter, sun for warmth.

Bibliography

1 Permaculture: A Designers' Manual
2 Designing and Maintaining your Edible Landscape Naturally
3 Introduction to Permaculture
4 Earth User's Guide to Permaculture
5 Global Gardener

Water Management

Chris Dixon

Why I include this session

Water management is fundamental to the development of sustainable systems at all scales, from garden to planet. The three essential criteria for the assessment of sustainability in any ecological system are water, soils and biodiversity, in that order.

Objective

To encourage a holistic understanding of water management.

Learning outcomes

By the end of this session students will be able to:

· Explain water management in a holistic way.

· Recognise water as essential to life and life processes.

· Recognise that primary water storage takes place in the soil and therefore attention needs to be given to the soil.

· Identify strategies for reducing run-off and increasing absorption.

Context

This general introduction can be taught early in a course. It can be followed by more detailed work such as reed beds, aquaculture, waste water recycling etc.

Duration

45 minutes

How I teach this session

Interactive talk

45 minutes

I generate the main points from the group and collate them on the board, or supervise as a group member collates.

Main points

· Water is essential to life.

· A holistic approach is vital.

· The failure of existing water management strategies is due to the fragmented approach: farmers, town planners, river authorities, water companies.

· Most water storage takes place in the soil, not in reservoirs etc: of the total water available on the planet, 11% is stored in soils as opposed to less than 0.5% in ponds, lakes, reservoirs and rivers. Generally within a few hours of the cessation of rain, run-off ceases and reservoirs and streams are fed by water from soil storage.

The problem

At present, the emphasis is placed on reservoirs, fed by run-off, as the main water storages. Ditching and drainage are often undertaken above such storages in order to increase run-off. Run-off is also increased by drainage of agricultural land and marshes. The opportunity for water absorption by soils is reduced by deforestation, by soil impaction from animals and machines, and by hard surfaces, such as buildings and roads, especially in towns.

All the above encourage removal of water from land-based systems as rapidly as possible; at best into reservoirs, at worst, and more likely, into rivers and thus the sea. Run-off also increases erosion. As the speed of flow increases so does the amount of debris which can be carried, and therefore the amount of materials lost from the land-based system. As designers we should observe and imitate natural processes which have evolved to ensure minimum run-off and maximum absorption in soils.

Solutions

Reducing run-off and increasing absorption can be achieved by:
- Reforestation in general and especially in upland areas – as a rule of thumb, the top third of hills.
- Continuous plant cover, especially in forests, eg selection forestry, non-felling on steep slopes.
- Contour fencing, with or without associated plantings.
- Deep mulch.
- Swaling, where rainfall is less than say 1 m per annum.
- Shallow diversion drains, where rainfall is greater than 1 m per annum, with a fall of 1:400 or less.
- Sensitive road siting.
- Non-drainage of upland marshes, and blocking of existing drains.
- Wetland regeneration, especially in uplands.
- Keyline, the Yeomans system, which may be appropriate in large scale design.

Multiple outputs

It should be noted that all the above are also anti-erosion strategies.

As designers we should also be concerned with increasing overall yield by ensuring that any plantings associated with water management contain economically valuable species, eg fuel, fodder, food, cash crops etc.

Urban areas

In towns and cities water absorption into soils may be neither appropriate nor possible. Increased attention needs to be given to:
- trapping water from roof and road surfaces.
- cleaning and recycling.
- downstream absorption and storage, in soils if possible, otherwise in reservoirs.

○ Link to ethics and principles

Earth Care: effective, sustainable management of water resources is essential to existing ecologies.
People Care: adequate supplies of clean water are essential to human settlements.

○ Further activities

- Sand tray or pit plus watering cans, to demonstrate swales, keylining, river formation, deltas etc.
- Rainfall gauges can be set up at the beginning of courses and monitored.

- Stream and river flow gauges can be devised and constructed on courses.
- Observation of water management in the locality, including: overflows, concreted areas, sewers, rivers, erosion sites, impacted soils, drainage systems.

Agenda 21 link

Chapters 13(B) & 18.

Further research

Water for every farm

Living Water, Alexanderson, O, Gateway, 1990

Living Water, 5 Holyrood Road, Edinburgh, EH8 8AE. Tel: (0131) 558 3313

Email: living.water@clan.com

Web: http://www.clan.com/environment/livingwater/

Aquaticum Treatment Wetlands, 49 Penbury Street, Worcester, WR3 7JD. Tel: (01905) 453815

Spiral Patterns

Rod Everett

○ Why I include this session

The observation of spiral patterns in nature and understanding their function can help us to develop an ability to use patterns in the design process. Spirals occur at atomic level through to the solar system. By understanding spirals we can change our thinking process away from the linear model to one that is more sustainable.

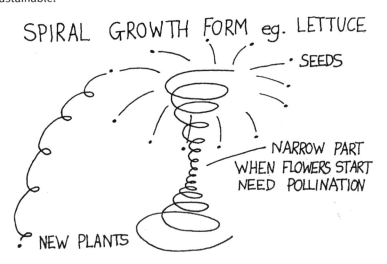

○ Objective

To use spiral patterns to stimulate the ability to integrate patterning in the design process.

○ Learning outcomes

By the end of this session, students will be able to:
· Identify spiral patterns in nature.
· Relate spiral pattern to function.
· Understand spirals by word, touch, sight and action.
· Begin to think in a non-linear manner.

○ Context

This fits well early on in the course and gets everyone participating actively.

○ Duration

45-90 minutes

○ *How I teach this session*

Observation exercise

Each student is given an object which relates to spirals. They are given three to five minutes to observe, then they discuss the pattern it shows and why it might be beneficial. Students are encouraged to consider the viscosity and hardness of the object and frictional forces.

Suitable objects are: shells, pea or passion fruit tendrils, rose or other flower head, cabbage cross section, seeding lettuce, sunflower head, seeding foxglove, water down a plug hole, spoon stirring water in a glass, pictures of a tornado or nuclear explosion, sheep's horn, picture of embryo developing, mushroom.

Discussion and presentation

Where else have students seen spiral patterns? Relate them to Overbeck Jet Model, Von Karman Trail and Ekman Spiral (Ref 1).

Thinking exercise

Think of designed objects that have used spirals, eg herb spiral, flow forms, corkscrew, turbines, propellers, high chimney with spiral vanes to catch wind, old brick wells. How can we learn from nature? (Ref 1, pp83-88)

Other points

· Business training uses the pattern of a continuous spiral pattern (Ref 2). It goes through an evolutionary/learning development, from dependent to independent to interdependent.

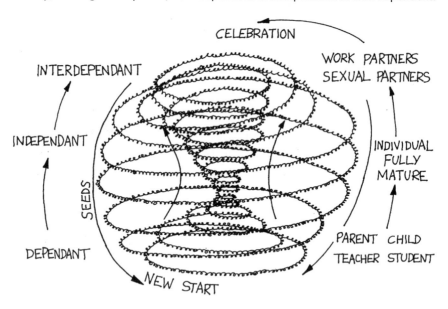

· Developing human potential and ecological potential has to go up one cycle at a time. At each cycle the element or combination of elements gains more knowledge, more fertility, and more ability. On reaching the celebratory climax, seeds are spread out to start the process again.

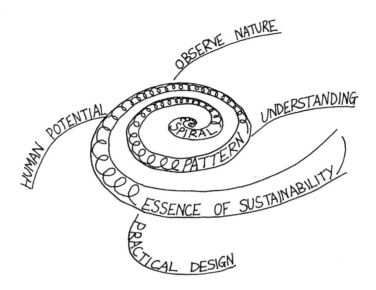

· Other places where spirals are used: energy around chakras, dowsing, peace dance, ziggurats, spiritual/spire/inspiration and ancient stone carvings, eg Newgrange.
· PMI and LETSystems create spirals of knowledge and wealth.

Spiral meditation

Start standing with feet apart, knees slightly bent and back straight. Hands are facing down to the ground. Imagine a spiral starting deep in the ground linking you strongly to the earth, learn from the earth. The spiral rises up to your centre to a point of stillness and great energy. Then it rises up linking to the surrounding energy of people, plants, air. At it's highest point high up in the sky there is celebration and joy at everything working in harmony. Then this is spread out as the spiral lowers, going out to the surrounding land and people and spreading energy out to whoever you wish, and then grounding the spiral back into the earth. This is repeated three or four times. The arms and whole body can join in so you can really feel the spiral energy!

○ Bibliography

1 Permaculture: A Designers' Manual. Chapter 4
2 The Seven Habits of Highly Effective People, Covey, S, Simon & Schuster, 1992

○ Further research

Patterns in Nature, Stevens, PS, Penguin, 1974

The Plant between Sun and Earth, Adams, G, & Wheeler, O, Shambala, 1982

Sensitive Chaos – The Creation of Flowing Forms in Water and Air, Schwenk, T, Rudolf Steiner Press, 1996

Living Water, Alexanderson, O, Gateway, 1990

Spiral Dance, Starhawk, Harper, 1989

Patterns in Design

Joanne Tippett

Why I include this session

The action of applying design to the landscape is one of creating patterns. I feel it is important to teach this area as a discrete topic in order to increase people's understanding of their effects on the landscape, and of ways in which those effects can be made more sustainable and congruent with ecological processes.

Objective

To introduce the theory of pattern understanding and show how this relates to the process of design.

Learning outcomes

By the end of the session, students will be able to:
· Describe patterns commonly found in ecological and self-organising systems.
· Show how these patterns can be used in design.
· Make reference to the disciplines involved in pattern understanding, and be able to conduct further research of their own.

Context

I teach a session dedicated to patterns at the end of the first week, directly before the review process. I teach it in the morning, as it is too dense a topic to leave to the vicissitudes of after-lunch or end-of-the-day tiredness. This two hour session is used to tie together many of the themes which have been covered through the week, and acts as a bridge into the design work of the second week.

Many of the ideas taught in the understanding of patterns are implicit in other permaculture principles, such as edge effect, stacking in space and time, zones, sectors and designing for surge and pulse. These patterns are made visible through the structure and methods of teaching, in the physical exercises and links made between the principles, and in the graphics used to represent the principles. Thus, when the principle of pattern application is taught, it acts as a reinforcement for many of the lessons learned during the week.

Understanding of patterns is deepened during the teaching of design process in the second week. In this design process, patterns are explored in terms of social organisation, landscape, interaction of processes and elements in design and in terms of how a design is elaborated on the landscape.

Duration

130 minutes including breaks

How I teach this session

I use a combination of exercises and observation, with examples from slides and the landscape around the classroom.

Resources required

Concrete area or other floor space you can draw on with chalk (or use large sheets of butcher's paper), coloured chalk, leaves, sticks, shells, flowers, etc. Example slides and projector, a cauliflower.

Whole group exercise

Stage 1

10 minutes

Pass around leaves, flowers, etc, and chalk. Ask people to start drawing the patterns they see in their objects on the concrete or butcher's paper – NOT to reproduce them exactly, but to do it quickly, identifying a pattern and repeating it. Demonstrate this.

Then ask the students to draw bigger and smaller patterns, whilst randomly moving about the drawing space. Ask participants to concentrate on the edges of theirs and other people's drawings, either by drawing their pattern into ones already drawn, or by drawing simultaneously with another person.

Quick discussion – stand up and look at it – ask "What do you see? What does it make you think of? Can you see any other permaculture principles in the drawing?"

Stage 2

10 minutes

Have students draw systems, eg orchards, animal housing, buildings, water catchment, into the previous pattern drawing, envisioning it as a map of an imaginary landscape.

Stage 3

15 minutes

Have the group move around the drawing, looking at the sketches and ideas. Quick discussion – how is this useful/interesting? Emphasize how patterns create edges and discuss how this can be used in design.

Remainder of the session

Walk

10 minutes

Look at patterns in landscape.

Group mind-mapping session

10 minutes

What are patterns?

Direct teaching with poster

15 minutes

Slides, case studies

20 minutes

Further discussion and questions

15 minutes

I use a cauliflower to teach self-similarity – ask people to describe its structure. Break off a piece and show how it is similar in structure to the whole cauliflower. From this piece, break off a smaller piece, showing how it too is similar to the whole. Discuss how this principle of self-similarity makes permaculture a relatively simple idea. At first, it can all seem rather complicated, but just as nature is self-similar at different levels of scale, in permaculture we learn that principles and patterns can be applied at many different levels of scale. Each time these are applied, the design will be different, as the principles are responsive to the unique site and situation.

Link to ethics and principles

I use the pattern session to tie together all of the other principles. During the session, I will walk around the classroom, pointing out the patterns and linking themes in the posters and group work mind maps which are up on the walls. I also use this session to consolidate the teaching about sustainability from the earlier session on The Natural Step model. (See pages 85-90.)

Further activities

The practical session after lunch on the same day, eg planting a forest garden edge, or building and planting a key hole bed, includes aspects of pattern application.

Further research

A Pattern Language
The Web of Life
A Pattern Language of Sustainability, Tippett, J, BA thesis at Lancaster University – unpublished, 1994. For web site address, contact holocene@juno.com
Fractals, The Patterns of Chaos, Discovering a New Aesthetic of Art, Science and Nature, Briggs, J, Thames and Hudson, 1992
The Tree of Knowledge, Maturana, H, & Varela, F, Shambhala, 1987

⇨ ## *Woodland*

Ben Law & Oak

○ ### *Why I include this session*

Trees are central to permaculture, and will play a large part in any sustainable culture. Additionally many students have a keen interest in them, so it is a session where students can 'show what they know'.

○ ### *Objective*

To explain the transaction of trees and woodland in their own right; and in relation to other life forms, species and land patterns and therefore their role in permaculture systems.

Learning outcome

By the end of this session, students will be able to:
- Understand the natural levels within woodland and different woodland types.
- Explain the environmental benefits of woodland.
- Apply for grants and be clear about the processes to establish new woodland.
- Design woodlands into permaculture systems.

Context

Woodland fits into the Design Course well if taught prior to 'Forest Gardening', as the natural levels of a woodland will be seen on the woodland walk and the concepts of multi-levels and utilising vertical space as a natural pattern should be becoming clear.

Duration

120 minutes

How I teach this session

Types of woodland

60 minutes

Taught in a slide show format or better still by a walk through a varied woodland, identifying different types of woodland and management patterns. If a walk is chosen this could take the whole hour. If not, a half hour walk should follow the slide show to reinforce the information, and as a stimulus for questions and discussion.

Brainstorm

20 minutes

Benefits of trees and woodland products.

Presentation

40 minutes

Present a permaculture woodland. This could be a case study of a real example or a visual planting design. Grant availability should be linked into either of these: what grants has the case study used or what is available for the visual planting? The session should end with handouts and contacts. Mentioning the grant system turns the possibility of planting a woodland or restoring a neglected coppice into a potential reality for students (Refs 1, 2 & 3).

Living examples

Leckhelm Wood, Scotland (Ref 4, p11)
Dun Beag, Scotland, Tel: (01700) 811 296
Prickly Nut Wood, Sussex (Ref 4, p110)
Keveral farm, Cornwall (Ref 4, p129)
Bioregional Development Group, Surrey, Tel: (020) 8773 2322
Regional Coppice Groups

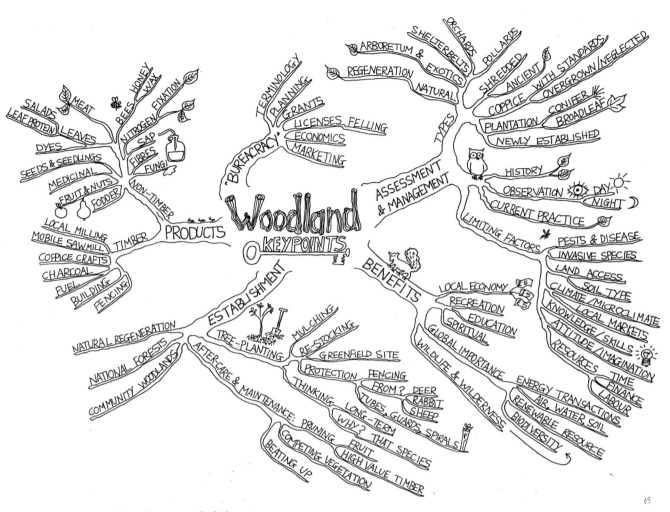

Further activities

The mind-map shows a range of possible activities and sessions that could be run on a Design Course concerning woodland.

Agenda 21 link

Chapter 11.

Bibliography

1 Forestry Authority, Tel: (01223) 314546 (ask for your Regional Office)
2 Small Woods Association, Tel: (01327) 361387
3 The Tree Council, 35 Belgrave Square, London, SW1X 8QN. Tel: (0120) 7235 8854
4 The Permaculture Plot

Further research

The Woodland Way, Law, B, Permanent Publications, 1998

Trees and Woodlands in the British Landscape, Rackham, O, Weidenfeld & Nicolson, 1976

Low Impact Development, Simon Fairlie, Jon Carpenter, 1996

Silviculture of Broadleaved Woodland, Evans, J, HMSO, 1984

Trees and Aftercare, Kiser, B, BTCV, 1978

Tools and Devices for Coppice Crafts, Lambert, F, CAT, 1987

Green Woodwork, Abbott, M, Guild of Master Craftsman Publications Ltd, 1989

Andy Goldsworthy, Goldsworthy, A, Viking, 1990

Woodlands, Brooks, A, BTCV, 1980

Woodland Crafts of Britain, Edlin, H, Batsford, 1947

Woodland in Permaculture, Whitefield, P, Self-published, 1992. Booklet

Caring for Small Woods, Broad, K, Earthscan, 1998

Farm Woodland Management, Blyth, J, et al, Farming Press, 2nd edition 1991

Planting Native Trees and Shrubs, Beckett, K & G, Jarrold, 1979

Practical Forestry for the Agent and Surveyor, Hart, C, Alan Sutton, 1991. The ultimate reference work for foresters

Spirit of the Trees (video)

Forestry Contractors Association, Tel: (01467) 651368

Green Wood Trust, Tel: (01952) 432769

Reforesting Scotland, Tel: (0151) 226 2496

Tree Spirit, Tel: (01299) 400 586

Bioregional Development Group, Tel: (020) 8773 2322

The Functions of Trees

Cathy Whitefield

Why I include this session

Trees are central to permaculture. This session allows us to look at trees in an overall way, rather than concentrating on one function or type of tree, such as orchards, windbreaks, coppice etc.

Objective

To give students a holistic introduction to the functions of trees.

Learning outcomes

By the end of the session students should have a subjective and objective appreciation of trees and their function in the biosphere.

Context

This can be taught at any time, but usually goes with the other woodland sessions.

Duration

30-45 minutes

How I teach this session

First I place a flipchart or small blackboard on the floor and ask the students to sit comfortably in a circle around it.

Visualisation

Leave a spell of silence between each of the following paragraphs. How long to leave varies from one paragraph to another, and according to the group you are working with. It may be in the range of 10 to 45 seconds.

· Think of a place you know where there are trees. It may be a woodland, urban park, orchard or any other place with trees. Imagine you're approaching the trees. Now you are walking between them, beneath their canopy.

· Feel the difference in the air quality between outside the area of trees and inside it: the wind, the temperature, the humidity.

· Listen for any sounds you can hear, perhaps birds, branches creaking in the wind, or the sounds of your feet meeting the woodland floor.

· Breath in deeply and take in the smells of the place.

· In your mind's eye look around. See the trees, the colours, shapes, textures and patterns of the leaves and trunks, see the other plants, and the quality of the light.

· Feel how being in this place affects you, your feelings and emotions. Be aware of changes in your body and breath which go with your changing emotions.

· In your own time, saying an inner goodbye and thank you to the trees, slowly walk out of that place, and come back to the room. Begin to take some deeper breaths, open your eyes, and have a stretch if you feel like it.

Brainstorm

Following directly on from the visualisation, with people remaining in the same positions, ask them to name the functions of trees. You may chose to keep solely to ecological functions, ie those which benefit the ecosystem or biosphere as a whole, or include those functions which are of particular interest for humans. (See mind-map below).

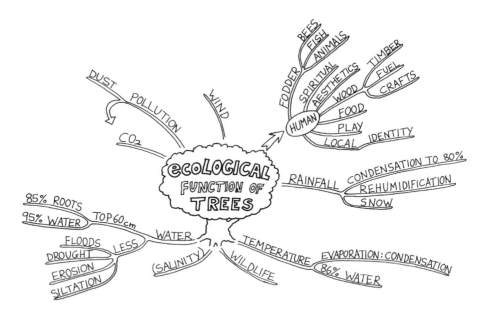

Talk

This is also best done without moving. Add any functions which may have been missed, and give some more detailed facts where necessary (Ref 2).

○ Agenda 21 link

Chapter 11(A).

○ Bibliography

1 Creative Visualization, Shakti Gawain, Eden Grove Editions, 1988
2 Permaculture: A Designers' Manual, Chapter 6

Native Trees

Rod Everett

Why I include this session

This encourages students to get to know local tree species, and shows how they can be integrated into permaculture design encouraging multi-functional use.

Objective

To stimulate an interest in native trees and their multi-functional use in permaculture design.

Learning outcomes

By the end of this session, students will be able to:

· Identify some local native trees.
· Relate the specific qualities that are useful.
· See the potential of using native trees in permaculture design.

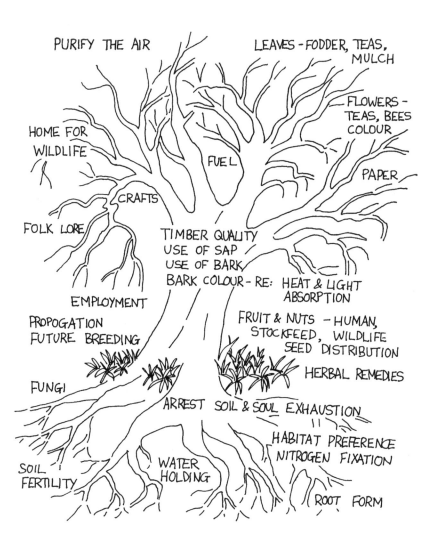

PURIFY THE AIR
LEAVES - FODDER, TEAS, MULCH
FLOWERS - TEAS, BEES COLOUR
HOME FOR WILDLIFE
FUEL
PAPER
CRAFTS
FOLK LORE
TIMBER QUALITY
USE OF SAP
USE OF BARK
BARK COLOUR - RE: HEAT & LIGHT ABSORPTION
EMPLOYMENT
PROPAGATION FUTURE BREEDING
FRUIT & NUTS - HUMAN, STOCKFEED, WILDLIFE SEED DISTRIBUTION
HERBAL REMEDIES
FUNGI
ARREST SOIL & SOUL EXHAUSTION
HABITAT PREFERENCE
NITROGEN FIXATION
SOIL FERTILITY
WATER HOLDING
ROOT FORM

Context

Middle of full Design Course.

Duration

As a walk outside: 90 minutes
As a slide show: 45 minutes

How I teach this session

There are two different techniques I have used for this session

1. Outside walk

To walk around woodlands identifying specific tree species and discussing their specific qualities as:

· A living tree: shape, form, habitat, preferences, shelter, hedgerow, fruits or nuts, leaves, flowers, wildlife associations, ability to withstand pollution, ability to coppice or pollard, potential for improving use by breeding, eg edible acorns.
· Timber qualities: strength, flexibility, rot resistance, grain quality, specific craft uses. Splitting logs of different wood types illustrates some qualities well.
· Folklore, herbal remedies, links to festivals and celebrations.

2. Slide show and discussion

Show slides of shape, form and leaf of specific tree species and discuss qualities as above. Include slides of woodland crafts.

Handouts

· Uses of native trees.
· Details of grants available.

Agenda 21 link

Chapter 11.

Further research

Woodlands, Brooks, A, BTCV, 1980
Field Guide to the Trees and Shrubs of Britain, Readers Digest, 1981
A Modern Herbal, Grieve, M, Penguin & Peregrine Books, 1980
In a Nutshell, Common Ground, 1989
The Sacred Tree, Kindred, G, self-published, 1995
Woodland Crafts of Britain, Edlin, H, Batsford, 1947
Sustainable Forestry, HMSO, 1994 UK Programme, CMD 2429
The Woodland Trust, Autumn Park, Dysart Road, Grantham, Lincolnshire, NG31 6LL. Tel: (01476) 58111
British Trust for Conservation Volunteers (BTCV), 36 St Mary's St, Wallingford, Oxon OX10 0EU.
Tel: (01491) 839766
Forestry Authority. Tel: (01223) 314546. Ask for your Regional Office
English Nature, Northminster House, Peterborough PE1 1UA. Tel: (01733) 455000
Contact your Local Authorities/Tree Wardens

Windbreaks

Patrick Whitefield

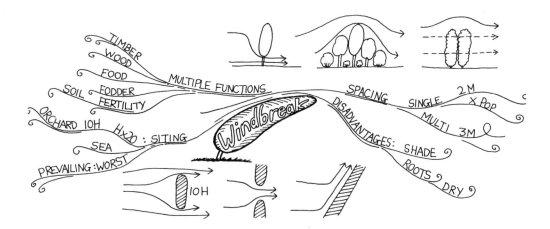

Why I include this session

It covers two permaculture specialities:

a) windbreaks creating special micro-climates

b) linear woodland giving plenty of edge between trees and open country, or between trees and water.

There are plenty of non-permaculture sources for information on conventional woodland, whether existing or new plantings, whereas windbreaks and shelterbelts have been largely ignored since the early 1960s. If I could only cover one topic under trees and woodland on a Design Course it would be this one.

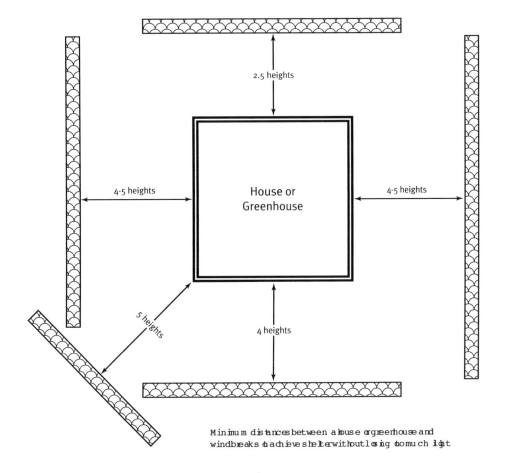

Minimum distances between a house or greenhouse and windbreaks to achieve shelter without losing too much light

○ ## Objective

To give a concise outline to the subject.

○ ## Learning outcomes

By the end of the session students will be able to:

- Explain some of the main concepts of windbreak design.
- With reference to their notes, approach the design of a windbreak with all the basic concepts to hand.
- Access more detailed information.

○ ## Context

Not crucial, but I usually include it with other woodland (Zone 4) stuff.

○ ## Duration

30 minutes

○ ## How I teach this session

Talk with sketches

I introduce the basic concepts of internal windbreak design, drawing them on the board as I go:

- Permeability.
- Making a solid wall into a more useful windbreak.
- Minimum length.
- Gaps and baffles.
- Orientation to prevailing wind.
- Importance of well-filled bottom.
- Profile: angled or vertical.
- Plant spacing.

Siting:

- Effectiveness of windbreak as multiple of height.
- Spacing of windbreaks for different purposes.
- Ideal distance from house or greenhouse.
- Prevailing wind often not the most damaging.
- Windbreaks near the sea; species choice.

Disadvantages:

- Shade.
- Root competition.
- Land take.
- Slow return.

Multiple outputs:

- Timber.
- Wood.
- Food.
- Fodder, including for fish.

- Soil fertility.
- Wildlife.
- Aesthetics.

Agenda 21 link

Chapter 11.

Further research

Shelterbelts and Windbreaks, Caborn, JM, Faber, 1965. The standard work

Trees for Shelter, Harriet, P, et al, (Eds), Technical Paper No 21, Forestry Commission, 1997. Includes: shelter for buildings, benefits to air quality, shelter for animals, shelter and wildlife, economics of shelter on farms

Woodland in Permaculture, Whitefield, P, self-published, 1992. Contains ideas on linear woodland

Farm Woodland Management, Blyth, J, et al, Farming Press, 1991. Contains some information on shelterbelts

Designing and Maintaining your Edible Landscape Naturally. Especially pp67-71 'Garden windbreaks'

Introduction to Permaculture. Especially pp45-50

Zone 1

Mark Fisher

Why I include this session

For urban dwellers – 80% of us in the UK – access to land is either non-existent or limited to gardens, allotments, a balcony or paved backyard. So Zone 1 has more significance for us than any other apart from Zone 0, our home.

As permaculturists, we have reached the decision to increasingly take personal responsibility for our existence. Growing some of our own food is probably the best example of this, and it is an activity primarily located in Zone 1. Thus the potential outcome from Zone 1 is an increasing self-reliance.

Objective

To show how to develop Zone 1 in relation to its scale and the resources available.

Learning outcomes

By the end of this session, students will be able to:
· Recognise that, whatever their circumstances, there are identifiable ways in which to develop productivity in their Zone 1.
· Begin to design personal food production systems that make use of the permaculture principles.

Context

It can be twinned with a follow-on topic such as Local Food Links, Local Agenda 21, or LETS. It is assumed that zonal and sector analysis will have already been covered.

Duration

60 minutes

How I teach this session

I allow the opportunity for personal discovery while presenting examples of food growing in Zone 1. These examples are most easily demonstrated with slides or line drawings and can include balconies, backyards, gardens and allotments. The group are encouraged to make connections with their own lives, both at present and in the future, and to reap the rewards of developing their Zone 1.

Opening review

I say the following before focusing on Zone 1. This can also be included in a handout.
· Zonal analysis looks at the energies used at a location or in social groups. This can mean all useable sources of energy, such as the activity of people, machines, products, waste materials and the careful use of hydrocarbon fuels.

· We use our energies in a measured way, matched with the activities associated with each zone. More of our energy will be used closer to our home, ie Zone 1, than would be used further away. We conserve our energies by thinking about the frequency of those activities, ie how often we need to do them, how far they are from our backdoor, and the time we have available to do the task. We also conserve energy by having multiple purpose in as many of our components as possible. Thus we have structures, materials and plants that allow more than one function, or close the circle of resource use.

Interactive talk to include the following points

· The basis of Zone 1 activities is high attention and intensity, and a strong interdependence with Zone 0.
· Because of the lack of access to broadscale land, there is a tendency to try to bring design components into a Zone 1 that may better fit in an outer zone. This can reduce the productivity of Zone 1.
· From starting on a small scale outside our back door we learn that we should only take on as much as we can successfully achieve.
· Identify what components are present and how they are interdependent, rather than give off-the-shelf designs.
· Indicate how designs can make activities supportive of each other, eg
  · composting organic waste to grow more food
  · collecting rain and grey water for irrigation
  · conservatory for extended growing season and solar house heating.
· Allotments:
  · may be some distance away
  · making it a personal space helps to bring it into Zone 1
  · consider growing food in combination with others, sharing work and produce
  · permaculture is often the spur to community food growing.
· For detailed food growing skills I recommend a local Organic Gardening course.
· I emphasise at the end of the session that every contribution in Zone 1 is valuable, from the window box on a balcony, the pot of herbs in the back yard, to the vegetable patch in the garden. This builds confidence in the land-poor and gives significance to their level of achievable scale.

Design exercise

· List components you have or would have in Zone 1.
· Which principles guide you in selecting them?
· What links can you make between them?
· How do they link in with other zones?

○ Link to ethics and principles

Significant principles include: starting near, maximising edge, optimum sizing or achievable scale, multiple use, giving everything a purpose, and stacking. The outcome of all Zone 1 activities will be increased self-reliance.

○ Further research

The Ornamental Kitchen Garden, Hamilton, G, BBC Books, 1990
Urban Permaculture
HDRA, Ryton Organic Gardens, Ryton on Dunsmore, Coventry CV8 3LG. Information on organic gardening and local organic gardening courses

Gardening

Patrick Whitefield

Why I include this session

Gardening has always had pride of place in permaculture courses. One of the central theories of permaculture is that we can feed ourselves much more efficiently from gardens than from farms, and a sustainable future would see farms replaced by gardens worldwide.

However this theory is by no means proven. Also to some extent the emphasis in permaculture courses is moving away from food production towards a more general look at sustainable living. Since the food we consume forms such a small part of our total ecological footprint, and what we can grow in our own gardens is but a part of that, gardening could be regarded as relatively unimportant.

That may be true mathematically, but humans are not solely mathematical beings. The most effective ways of reducing our ecological footprint tend to be negative: not doing something we would otherwise do, like flying, driving a car or consuming certain products. Gardening, on the other hand, is something positive and active we can do to make the world a better place. It inspires people.

Courses on 'how to be green' tend not to happen due to lack of interest, whereas permaculture courses and publications are getting more popular all the time. Gardening is an essential part of a package, which also includes the 'biggies' like turning down the thermostat.

Objective

There's no point in trying to teach people general gardening skills on a Design Course. There's not enough time to teach a significant amount, and gardening instruction is widely available elsewhere. Instead I concentrate on the distinctive contribution which permaculture can make to gardening. In Gardening 1 and 2, I do this by looking at how the permaculture principles can be applied in the garden, especially the key planning tools. This information is reinforced by the more experiential Placement Exercise, and illustrated in the slide show, An Urban Garden.

In the Mulching session I do go into more practical detail. This is one gardening method which is particularly characteristic of permaculture, and not often taught comprehensively elsewhere. In this session I aim to give students enough information that they can actually go away and start mulching.

Context

On some of my Design Courses the first two days consist of an introductory weekend, which can be taken on its own or as part of the Design Course. This weekend always contains Gardening 1 and the slide show on Mulching. Gardening 2 comes early in the main body of the course.

On other Design Courses I usually place Gardening 1 immediately after the Principles. It serves as an example of the principles in action, and one which most students can relate to their own lives, thus reinforcing the learning of the principles. The slide show, An Urban Garden, is usually shown the same evening. Gardening 2 and Mulching usually come on Day 4, after Zone 0, which is the main theme of Day 3. (See my 'Example Timetable' on page 59).

The Placement Exercise can be done at any time after Gardening 1, but it's best placed in an afternoon or alongside a solid information session to lighten things up.

Duration

Gardening 1	45min
Gardening 2	45min
Mulching	30min
An Urban Garden	45min
Placement Exercise	30min

How I teach this session

Gardening 1

Learning outcomes

By the end of the sessions students will be able to:

· State how the key planning tools can be used in garden design.

· See how this can be done in one or more actual gardens.

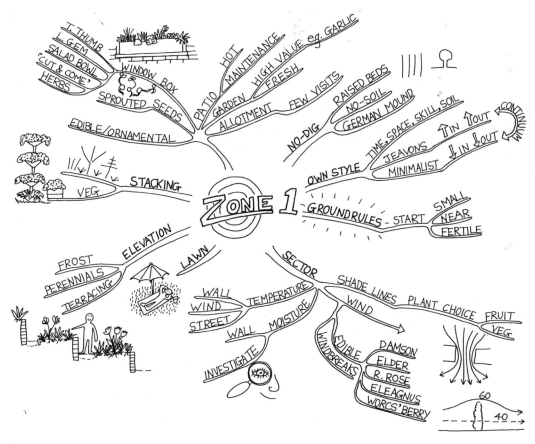

Talk with sketches

35 minutes

With a really interactive group this session can easily over-run, which is fine.

· Zone
  · the Kitchen Window Rule, the Gardener's Shadow.
  · edible/ornamental growing.
  · crops for different situations: indoors, window box/balcony, patio, home garden, allotment; criteria for choosing crops and examples.
· Sector
  · the Twelve Month Rule.

- shade: making a shade map, how to select plants for sunny and shady situations.
- wind: importance of shelter; wind tunnels; permeable windbreaks; edible windbreaks, pros and cons.
- moisture: the base of walls, investigating wet areas.
- temperature: heat reflection and storage, thermal mass, wind chill, sun traps.
- Elevation
  - terracing: pros and cons; two methods.
  - perennials for steep slopes (see slide show, Useful Plants, in the 'Forest Gardening' session, pages 246-250).
  - frost pockets.

Discussion

10 minutes

"Please get into groups of three and discuss how you could use these ideas in your own gardens. Make sure that there is no group consisting entirely of people who haven't got a garden."

   This often leads to one person drawing a map of their garden and telling the other two about it in detail. I don't try to prevent this, but at the end of the session I make the point that anyone who didn't get to talk about their place this time should do so in a future discussion session.

Gardening 2

Learning outcomes

By the end of the sessions students will be able to:
- State the possible uses of stacking, no-dig methods and lawns in permaculture gardens.
- Consider what kind of gardening style may suit them.
- State the three golden rules.
- See how the above may be applicable to one or more actual gardeners.

Talk with sketches and slides

30 minutes
- Stacking:
  - the importance of vertical space where horizontal space is limited.
  - fruit.
  - vegetables.
- No-dig:
  - raised beds
  - straight or keyhole, the myth and the reality.
  - gardening without soil: five slides from CAT.
  - German mound, a compromise for poorly drained or heavy soils.
- Lawns, pro and con:
  - used lawns and cosmetic lawns.
  - widening the concept of yield.
- Find your own style:
  - double-dig intensive method (Jeavons), the extreme high-input/high-output style.
  - the minimalist garden (edible perennials), the opposite extreme.
  - pros and cons of each.
  - find your place on the continuum.
- Three Golden Rules:
  - Start small – and stay small.

· Start near home.
· Start on the most fertile soil.

Whatever I may have to drop due to lack of time, I always include the Three Golden Rules.

Discussion

15 minutes

As above in Gardening 1.

Mulching

Learning outcomes

By the end of the session students should be able to:

· explain the functions of mulch, its advantages and disadvantages.
· mulch to help establish trees and hedges.
· use mulch in the garden, including the classic three-layer grow-through mulch.

Slide show

25-35 minutes

· Introduction, mainly brainstorm:
  · what is mulch?
  · the functions of mulch.
  · three kinds of mulch: clearance, grow-through and maintenance.
· 22 slides
  · 10 showing mulching of trees (if time is short these slides, which take about 10 minutes to show, may be postponed to the Trees and Woodland slide show later in the course).
  · 8 slides showing clearance and grow-through mulches in the garden, including a full demonstration of making the classic permaculture grow-through mulch.
  · 4 slides showing maintenance mulch and sources of materials.

An Urban Garden

Learning outcomes

By the end of the session students will be able to:

· See how much can be done in a situation which might be thought of as unpromising.
· Be empowered by this
· Be aware of some design features and techniques which can be used in a small garden.

Slide show

45 minutes

This is an inspiring example of what can be done in a small space by people with little or no gardening experience. The main subject is Michael and Julia Guerra's garden, where they grow a large proportion of their own food in a space many people would dismiss as too small. It also illustrates design in practice.

· 27 slides of Julia and Michael's garden, showing:
  · its development through time.
  · design for limited space, including stacking.
  · sectoring, especially with regard to sun and shade.
  · diversity.
  · use of local resources.
  · fitting into the social environment.

- how a productive garden can be beautiful too.
- 8 slides of other gardens, showing:
  - special micro-climates.
  - stacking.

Placement Exercise

Learning outcomes

By the end of the session students will:
- Have had some experience of using the key planning tools in practice.
- Be able to approach the relative placement of elements within a design with some experience of the process.

Activity

30 minutes

I divide the students into groups of three or four and give each group:
- a base map of an actual garden, including adjacent vegetation and buildings.
- a list of the features to be fitted into the garden, with approximate dimensions, eg 'salad and herb bed, 4 square metres'.

I then ask them to place the features within the garden, giving due attention to zoning, sectoring, plant-plant interactions and network analysis.

When they've finished I draw on the board what was actually done on the ground, giving reasons. I don't ask the students to present their plans, but point out that there is no right answer, and their solution may be equally valid.

○ *Living examples*

Many examples, including the Guerra's, are listed in The Permaculture Plot, with addresses, phone numbers and visiting arrangements.

HDRA (Ryton Gardens and one in Kent), CAT and Earthward all have demonstration gardens.

○ *Further research*

The Permaculture Garden

Designing and Maintaining Your Edible Landscape Naturally

How to Make a Forest Garden

The Indoor Kitchen Garden, Spoczynska, J O I, Bloomsbury, 1988. Over-optimistic on indoor light levels, but good for balconies etc

The Vegetable Garden Displayed, Larkcom, J, Royal Horticultural Society, 1992

The Salad Garden, Larkcom, J, Windward, 1984

Vegetables for Small Gardens & Plants for Small Gardens (two books), Larkcom, J, Hamlyn, 1995

Practical Mulching, Whitefield, P, self-published, 1992

Organic Gardening, Magazine, PO Box 4, Wiveliscombe, Taunton, Somerset TA4 2QY

Permaculture Magazine, Permanent Publications

HDRA (Henry Doubleday Research Association), Ryton-on-Dunsmore, Coventry CV8 3LG. Tel: {02476) 303517

CAT (Centre for Alternative Technology), Machynlleth, Powys SY20 9AX. Tel: (01654) 702400

THRIVE (formally called Horticultural Therapy), The Geoffery Udall Building, Trunkwell Park, Beech Hill, Reading RG7 2AT. Provide a comprehensive information service on all aspects of gardening as therapy

Pests and Predators

Nancy Woodhead

A predator is anything which reduces yield, including weeds, pests, poisons, pollution and anything which breeds fear.

Why I include this session

Because its an ideal example of do nothing or low intervention strategies in practice.

Objective

To describe and demonstrate a range of situations where loss of yield can be counteracted by good design and appropriate action.

Learning outcomes

By the end of this session students will be able to:
· Name a range of low intervention techniques.
· Visualise a positive solution to any pest problem.

Context

Early on.

Duration

In the classroom: 90 minutes
Garden demonstration: 60 minutes

How I teach this session

Opening exercise
20 minutes
I ask students to work in pairs to draw up a list of problems which have caused a loss of yield in their system. This can include slugs and snails, aphids, cats, bank managers, other bureaucrats, couch grass, dandelions, drought, waterlogging, New Zealand flatworms, etc.

Guided discussion
50 minutes
Put the list on one side and with the whole group, develop the following strategies:

The permaculture approach
Permaculture seeks to:
0 View predators as assets.
1 Absorb predation by offering no single target.
2 Accept some loss as natural.
3 Deter predators by minimum intervention.
4 Destroy predators... in that order.

Cascades of intervention

Minimum intervention for maximum productivity (a pattern)

· The cascade of intervention states that strategies should be chosen on the scale:

    0 Do nothing.

    1 Use biological intervention.

    2 Use physical/mechanical intervention.

    3 Use chemical intervention.

    (4 Use nuclear intervention).

· The first choice, do nothing, allows low-input strategies to prevail. If a system is whole and stability has been designed into it, there will be a balancing reaction to the predator, which will solve the problem. Eg a biodiverse cropping area will support a healthy insect population, including hoverfly larvae, ladybirds, ichumen wasps and so on, which will soon disperse a population of aphids,

· Biological intervention means using living organisms to control pests, eg ducks against slugs.

· Chemical intervention is sometimes a requirement, eg rat poison.

Strategies to use in the garden:

· Cultivate a healthy soil: soil building, composting.

· Select trouble-free plants: prefer locally proven varieties, avoid hybrids.

· Feed regularly: use plant fertilisers and soil conditioning.

· Rotate crops to discourage build-up of pest populations.

· Encourage natural predators: build ponds, create habitats for birds and insects.

· Encourage diversity.

Least toxic pest control examples:

· Cultural: sanitation, tillage, timing of planting, companion planting, crop rotation

· Physical: handpicking, traps, pruning, barriers, mulch.

· Biological: finding out about life-cycles, weak points and natural predators; encouraging beneficial species; bacterial insecticides; pheromone traps.

Thinking exercise

20 minutes

At this point I return to the original list and ask students to devise some possible solutions to deal with the problems already identified. I also ask participants about strategies they may have employed previously. I emphasise that pattern-thinking gives a strategy to make choices between different responses to predators.

Demonstration garden activity

60 minutes

For example, show nettles in a corner of the garden, and discuss how they can be employed; describe the use and placement of lacewing hotels, ponds with ducks, frogs, newts etc.

○ Living examples

See The Permaculture Plot and 'Light on the Horizon', Issue 13, Permaculture Magazine, Permanent Publications, 1996

Rubha Phoil, Armadail Pier, Ardvasar, Sleat, Isle of Skye, IV45 8RSS. See The Permaculture Plot

Ryton Organic Gardens, Coventry, CV8 3L. Tel: (02476) 303517. Also, 'Pest Control Without Poisons', Issue 16, Permaculture Magazine, Permanent Publications, 1997

○ *Agenda 21 link*

Chapter 14(l).

○ *Further research*

Healthy Fruit & Vegetables, Pears, P, & Sherman, B, HDRA/Search Press, 1991

Pests, how to control them on fruit and vegetables, Pears, P, & Sherman, B, HDRA/Search Press, 1992

Weeds, how to control and love them, Readman, J, HDRA/Search Press, 1991

Soil Care and Management, Readman, J, HDRA/Search Press, 1991

Companion Planting Chart, Littlewood, M, Self-published, 1997

Companion Planting, Frank, G, Thorsons, 1983

Bob Flowerdew's Complete Book of Companion Gardening, Flowerdew, B, Kyle Cathie Ltd, 1995

Collins Guide to Pests, Diseases and Disorders of Garden Plants, Buczacki & Harris, Harper Collins, 1998

Designing and Maintaining your Edible Landscape Naturally

Forest Gardening (3 sessions)

Patrick Whitefield

Why I include these sessions

Because:

a) Students may want to grow a forest garden;

b) It's a good context for a detailed look at some permaculture principles in practice.

c) The Fruit Placement Exercise is an introduction to the kind of thinking required in the design process.

Context

The three sessions described below are best done within a day or two of each other, in any order.

Objectives

a) To give a brief introduction to the information and skills needed to grow a forest garden, including reasons for choosing to grow this kind of garden and warnings about the most common pitfalls.

b) To illustrate permaculture by means of the example of forest gardening.

Duration

Main Forest Garden Session: 45 minutes
Fruit Placement Exercise: 20-30 minutes
Slide Show, Useful Plants: 35 minutes

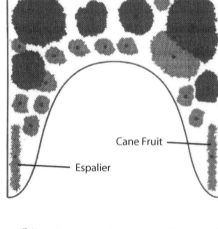

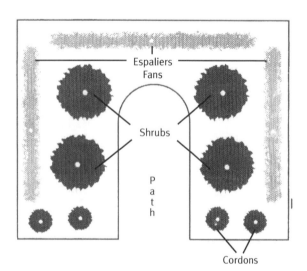

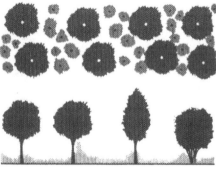

How I teach these sessions

Main Forest Garden Session

Learning outcomes

By the end of this session students will be able to:

· Decide whether they want to grow a forest garden.

· Consider whether some aspects of forest gardens can be included in other kinds of gardens.

· Outline the ecology of multi-layer perennial systems.

· List the criteria for choosing plants for a forest garden.

· Choose a suitable layout for a specific site.

· Explain the principles of designing a planting plan.

· Avoid common pitfalls of forest garden design.

Duration

45 minutes

How I teach this session

Slide show

20 minutes

Approximately 20 slides illustrating:

· The effect of different canopy species on lower layers.

· The effect of different spacing of canopy trees on light levels to lower layers.

· Succession.

· Edge.

· Annual growth cycles.

· Different kinds of forest garden.

· Maintenance.

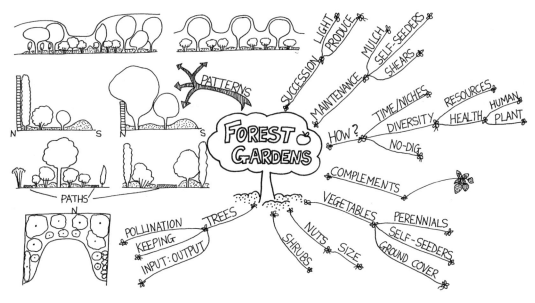

Talk

25 minutes

Covering:

· How a forest garden works: annual cycles, ecological niches, diversity, no-dig, perennials.

· How it complements an annual vegetable garden.

· Vegetable layer: how to choose plants, plant spacing.

- Shrub layer: how to choose plants.
- Nuts: possible plants, squirrels.
- Tree layer: how to choose plants, pollination, ripening times and keeping qualities, tree spacing.
- Layout: the importance of edge, 6 basic patterns (See illustrations XXX, XXX and XXX)
- Succession
- Maintenance

Fruit Placement Exercise

Learning outcomes

By the end of this session students will have had an experience of part of the process of designing a forest garden.

Duration

20-30 minutes

How I teach this session

This is based on the garden design illustrated on pp143 & 145 of my book How to Make a Forest Garden (Ref 1). (See illustrations below.) The task is to place the required trees relative to the environment and to each other.

Materials required for this exercise

Base map of the garden showing area where the trees are to go; circles of card representing the five main trees, three apples, one plum and one cherry plum, at mature size; printed sheet giving relevant characteristics of each tree variety.

Running the exercise

I divide the students into groups of three to five, according to the total number of students, and explain the task. I give all relevant information about the site verbally. I point out that this is only part of the design process; other decisions come before and after this task. There is no time limit. When all the groups have come to a decision, I show them my solution and give my reasons, pointing out that this is not necessarily more right than theirs.

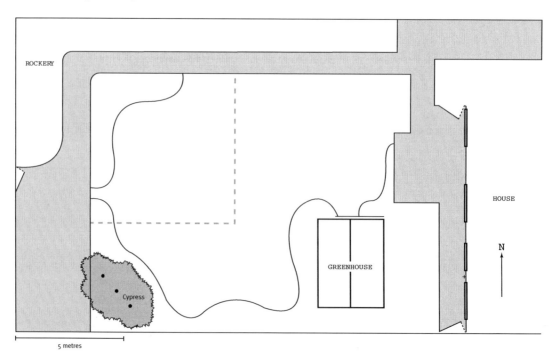

ROCKERY

HOUSE

GREENHOUSE

Cypress

N

5 metres

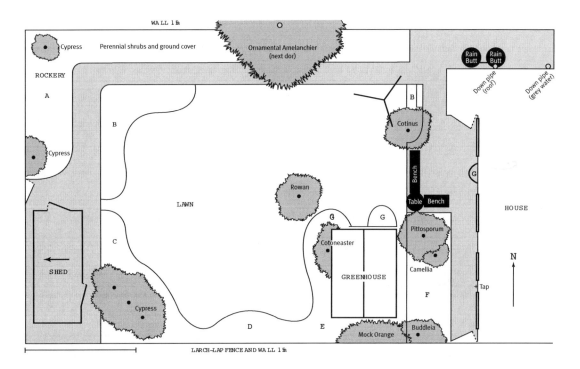

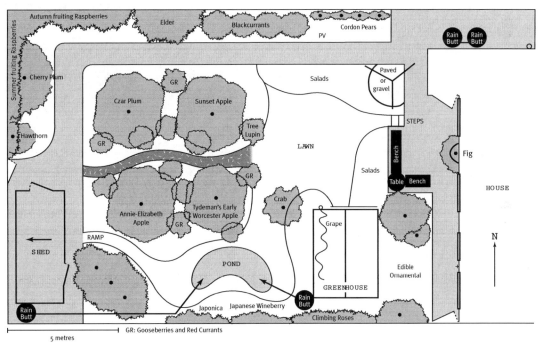

Slide show, Useful Plants

Learning outcomes

By the end of this session students will be able to approach the selection of forest garden plants with some idea of the range which is available.

Duration

35 minutes

How I teach this session

Slide show

Approximately 30 slides, of which approximately five show trees and shrubs, and 25 herbaceous plants, mostly perennial and self-seeding vegetables. As well as introducing a number of specific plants, this slide show illustrates the following points:

· The choice between low-input/low-output and high-input/high-output species.
· Ground covers.
· Native edibles.
· The importance of taste.
· Edible ornamentals.

Living examples

Many forest gardens, including Robert Hart's, are listed in The Permaculture Plot, with addresses, phone numbers and visiting arrangements.

Agenda 21 link

Chapter 14.

Bibliography

1 How to Make a Forest Garden

Further research

The Forest Garden, Hart, R, Institute for Social Inventions, 1991. A concise account of Robert Hart's pioneering work
Forest Gardening, Hart, R, Green Books. Robert Hart's personal testament, it adds a little practical information to the booklet above
Fruit, Baker, H, Royal Horticultural Society/Mitchell Beazley, 1992. A beginner's guide to growing fruit at home, covers all fruit and nuts which can be grown in Britain
The Fruit Garden Displayed, Baker, H, Royal Horticultural Society/Cassel, 1991. More detailed information on a narrower range
'Growing Nuts in Britain', Brown, R, Pts 1&2, in Permaculture Magazine, Nos 15 & 16, 1997
Food for Free, Mabey, R, Collins, 1989. Wild food plants
Agroforestry News, The Agroforestry Research Trust, periodical
Directory of Apple Cultivars, Crawford, M, Agroforestry Research Trust, 1994
Walnuts, Chestnuts, and Hazelnuts, Crawford, M, Agroforestry Research Trust, 1995-6. Three booklets, consisting of articles reprinted from Agroforestry News
Plants for a Future, Fern, K, Permanent Publications, 1997. Chapters 2, 3, 5 & 10
Designing and Maintaining Your Edible Landscape Naturally

Designing Systems for Animals

Lyn Dixon

Why I include this session

· Because of demand from course participants.
· It is an excellent way of showing a practical application of permaculture principles.
· The design of sustainable systems for animals is a pattern that can be applied to all species, including humans.

Objective

To enable people to design systems for animals which care for the physical and psychological well-being of both the animals and their keepers whilst creating a sustainable and healthy environment for all.

Learning outcomes

By the end of this session students will be able to:
· Realise that they can design systems for animals by appreciating the individual behaviour and needs of each species.
· Explain why observation and knowledge of each species is necessary before designing the system.
· Create a specific design from a general pattern shaped by the ethics and principles.
· Recognise that ideally the system should be up and running before the species is introduced.

Context

Middle of the course.

Duration

120 minutes

How I teach this session

Talk

30 minutes

Explanation of how to design systems for animals by reference to the ethics, principles and zoning. By doing this in a theory situation for more than one species with animals as different as, for example, horses and ducks, we can see how a pattern emerges that can be applied to all species.

Walk

30 minutes

Look at existing animal systems. The opportunity to compare sustainable and non-sustainable systems would be excellent.

Discussion

20 minutes

In small groups we discuss what we've seen, thought and felt, reporting back to the main group at the end.

Design exercise

20 minutes

A design for a site involving animals and wild ideas! The site can be hypothetical, a plan being drawn up beforehand and handed out, or an actual site, plans of which may be brought by course participants. Special consideration should be given to the physical and psychological needs of the livestock and their interaction with the environment, keeping in mind such principles as The Problem Is The Solution, and Work With Nature Not Against It. This can be done in small groups again.

Design exercise report-back

20 minutes

Groups report back with the teacher collating the information on the board.

Handouts

I provide information on things which cannot easily be arrived at through observation or discussion, eg plants which accumulate minerals (Ref 1).

Link to ethics and principles

I do this a great deal. Use them to check out your design and routine. They are a succinct way to get to the nub of things. Be adaptable: if it's not working, fix it. It's alright to have a plan, but don't fall in love with it!

Living examples

I would recommend a visit to our holding Tir Penrhos Isaf where the principles of permaculture and the teachings of Monty Roberts (Ref 2) are being combined in the implementation of our design of a sustainable and harmonious habitat for horses and more.

I would also recommend a visit to Matt and Jan Dunwell's farm, Ragman's Lane. This would be of particular interest to anyone contemplating a commercial venture in the production and sale of organic meat. Jan and Matt incorporate a number of additional, viable enterprises on their farm.

Both places hold courses and are listed in The Permaculture Plot.

Further activities

- · I often ask participants to bring a plan of their own land for a group design.
- · Designing the management of pasture and water.
- · Slide shows to show mistakes, solutions and successes.
- · Good items often come up during discussions in the tea breaks. Setting a topic and reporting back after breaks can therefore be very rewarding.

Agenda 21 link

Chapter 14.

Bibliography

1 Designing and Maintaining Your Edible Landscape Naturally
2 The Man Who Listens to Horses, Roberts, M, Hutchinson, 1996

○ ## *Further research*

The Horse's Mind, Rees, L, & Stanley, P, Ebury Press, 1993

Goat Husbandry, McKenzie, D, Faber & Faber, 1996

Zoning

00   Self and inner self – health and peace of mind – with animals, their health and peace of mind.

0   House – stables and outbuildings.

1   Food and fodder including cut and carry crops, eg comfrey and alfalfa, and possibly areas of fodder beet and grain. Also small paddock.

2   Orchards, including fodder trees and succouring trees, all acting as shade, shelter etc. Stack in hens and geese.

3   Main crops including permanent pasture and hay.

4   Fuel and forage, forest and its produce, hill grazing left low in nitrogen to provide good summer grazing for fat ponies or milkers being dried off etc.

5   Wilderness, where we can observe the evolution of natural systems. We can steer some areas by occasional grazing.

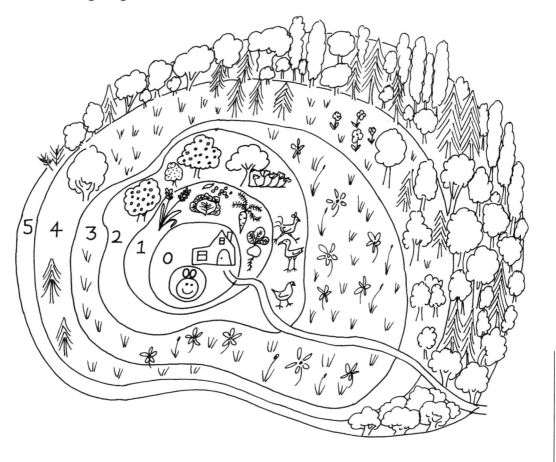

Farming/Zone 3

Patrick Whitefield

Why I include this session

Only a small proportion of the students on any Design Course are likely to be farmers, present or future. But all of us eat. Although a permaculture vision of the future may be of food being grown predominantly in gardens rather than on farms, in the short and medium term most of it, especially grains, will continue to come from farms. As responsible consumers and citizens it behoves us to know something about how this food is currently produced, and how it may be produced in a more permacultural way.

Also, the various options for grain production best exhibit the range of different ways permaculture principles can be applied to one activity, from the slightly permacultural to the extremely so.

Objective

To give students an idea of how permaculture may be applied to broadscale agriculture, and to deepen their understanding of permaculture generally.

Learning outcomes

By the end of the session students will be able to:
· State some of the characteristics of conventional farming.
· Explain the difference between the permaculture and organic approaches.
· Give some characteristics of the bicrop Bon Fils, and domestic prairie methods of grain growing.
· State some of the pros and cons of agroforestry in a British context.
They should also have had the opportunity to:
· Discuss any points of interest to them on agroforestry in particular and farming in general.

Context

This session usually takes place on Day 4 of the course. On Day 2 we will have done the exercise 'What Do I Eat?' (See pages 262-263). The programme for Day 4 goes like this:
· First session, Soil   60 minutes
· Break
· Second session, Farming   120 minutes
· Lunch
· Third session, Soil Practical   90 minutes
· The remainder of the day varies
At some other point in the course we cover Direct Marketing/Community Supported Agriculture.

Duration

120 minutes

How I teach this session

Part 1: Slides, general

30 minutes

24 slides, illustrating:

· Soil erosion on British farms.
· Biculture, both cereals and brassicas grown in clover (Ref 1).
· Bon Fils (Ref 2).
· Foggage.
· Appropriate breeds of animals.
· Alternative animal feeds.
· Agroforestry, slivoarable and silvopastoral, at three British sites.

Part 2: Talk, grain growing

40 minutes +

A comparison of six ways of growing grain

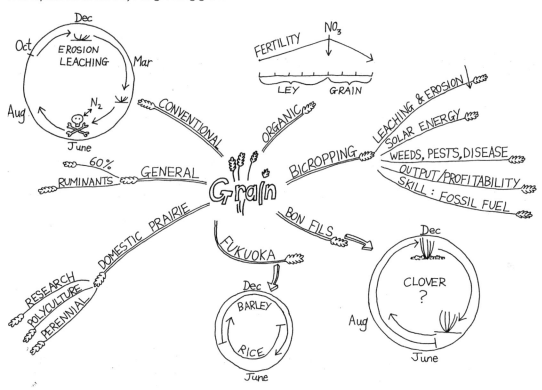

1 Conventional. Draw annual cycle and explain.

2 Organic. Draw rotation and explain (Ref 3, 4 & 5).

3 Biculture. Outline work of Bob Clements et al at Institute of Grassland & Environmental Research; note beneficial relationships and implications for yield/multiple output and profitability (Ref 6).

4 Bon Fils. Draw annual cycle and explain; note unresolved problem of controlling vigour of clover (Ref 2).

5 Fukuoka. Draw annual cycle and give brief outline (Ref 7).

6 Domestic prairie. Outline work of Wes Jackson et al at Land Institute; note long-term nature of work; probably the ideal solution for grain growing (Refs 8 & 9).

Other points:

· 60% of present grain harvest is fed to animals.
· Eating low down the food chain.

· The importance of ruminants in sustainable grain production ie sheep, cattle and goats digest cellulose, which gives them a unique role within ecosystems.

· Sustainable systems substitute skill for fossil fuel energy.

Part 3: Discussion, agroforestry

Time variable.

Give out handout listing pros and cons of agroforestry and ask students to read it. When they have finished, ask for comments and reflections. The discussion is often wide-ranging and need not be confined to agroforestry (Ref 10).

Agroforestry

Agroforestry means growing both trees and field crops, ie arable or pasture, on the same piece of ground at the same time. A traditional orchard, grazed by sheep and cattle is one example.

The advantages of agroforestry are:

· Better use of resources by plants with diverse size and structure, eg reduced competition for water due to different root depths.

· Increased biodiversity, giving a more stable and healthier agroecosystem.

· Increased yield of field crops due to manuring effect of leaf fall from trees.

· Protection of arable or grazing animals from wind.

· Shade for animals.

· All leading to: greater overall yield than if either component was grown alone.

· Reduction in soil erosion by wind or water.

· Lock-up of carbon by growing trees, reducing the greenhouse effect.

· Diversity of output insures against fluctuations in prices or weather.

· Tree products can provide raw materials for local crafts or industry and on-farm use, making local economy more self-reliant.

· Better habitat for wildlife and humans.

Its disadvantages are:

· Most farmers are unfamiliar with growing trees.

· Farming operations are more intricate.

· Income from the tree component is not immediate.

· This income may be low if value is not added on the farm.

○ Living examples

See references 11, 12 and 13.

○ Agenda 21 link

Chapters 10 & 14.

Bibliography

1 'Clover and Cereals – Low Input Bi-Cropping', Clements, B & Donaldson, G, Farming and Conservation, Vol 3, No 4, 1997

2 The Harmonious Wheatsmith, Moodie, M, self-published, 1991. Describes the Bon Fils method.

3 Organic Farming, Lampkin, N, Farming Press, 1994

4 Elm Farm Research Centre, Hamstead Marshall, Nr Newbury, Berkshire, RG15 0HR. Tel: (01488) 658298. Organic farming

5 Soil Association, Bristol House, 40-56 Victoria Street, Bristol, BS1 6BY. Tel: (0117) 929 0661. Email: soilassoc@gn.apc.org

6 Institute of Grassland & Environmental Research, North Wyke, Okehampton, Devon EX20 2SB. This is the address to contact Dr Bob Clements

7 The One-Straw Revolution

8 Farming in Nature's Image, Soule, JD, & Piper, J K, Island Press, 1992. Describes the work of the Land Institute

9 The Land Institute, 2440 East Water Well Road, Salina, KS 67401, USA

10 Agroforestry Forum, research journal, published by School of Agricultural and Forest Sciences, University of Wales, Bangor, Gwynedd LL57 2UW

11 'Broadscale Permaculture Realities at Ragmans Lane Farm', Dunwell, M, Permaculture Magazine, Issue 10, 1995. A mainly grassland farm, developing permaculture

12 'Profit from Mixed Crop', Stebbins, K, Farmers Weekly, 4 Sept 1992. An experiment in farm-scale agroforestry

13 'Trees Among the Wheat', Whitefield, P, Permaculture Magazine, Issue 11, 1995. A newly-established agroforestry farm

Further research

Bio-Dynamic Farming Practice, Sattler, F, Wistinghausen, F & E, Bio-dynamic Agricultural Association, 1992

Agroecology, the science of sustainable agriculture, 2nd edn, Altieri, M, Intermediate Technology Publications, 1995

Bio-Dynamic Agricultural Association, Goethean House, Woodman Lane, Clent Stourbridge, West Midlands DY9 9PX

Sustain, 94 White Lion Street, London N1 NPF. Tel: (020) 7837 1228

Aquaculture

Matt Dunwell

Why I include this session

Aquaculture is often overlooked in the UK. It is considered
to be only relevant to warmer countries like Australia.
However, as the UK climate becomes dryer and more
erratic we need to start looking at systems to hold water
in the landscape, and at how to use them productively.
Aquaculture systems are the most productive growing
medium even in temperate climates.

Objectives

To explain the role of aquaculture systems in the UK; to explain methods of construction and utilisation
with a view to encouraging aquaculture as a design priority. To encourage people to see boggy or
poorly drained land as a resource not a hindrance.

Learning outcomes

By the end of the session students will be able to:
· Design simple aquaculture systems.
· Understand the role of edge and micro-climate.
· Observe productive aquaculture systems and systems that can be improved.
· See potential for water catchment in the landscape.

Context

Aquaculture should be covered after ethics, principles and design sessions.

In design terms, pond sites have tighter constraints than house sites. They include aspect, slope, soil
type and rainfall. So water storage should be considered first, and designed into the landscape before
housing and settlement patterns. Therefore aquaculture needs to be considered early in the broadscale
design process.

Duration

50 minutes

How I teach this session

Talk – basic principles
25 minutes
· Look at hydrogen bonding as a way of explaining the properties of water.
· We can expect 4-10 times the yield of protein from water as from land: fish are cold blooded, do not
  have to support their body weight etc.
· Natural systems are breaking down, eg North Sea. Fish from seas and fish farms are increasingly
  polluted. We can get sustainable fish production in ponds: carp, maybe trout with enough water flow.

- Check water catchment is sufficient, and whether it's year-round or seasonal.
- Check catchment area for erosion, which will silt up the pond, and contaminants, eg industrial, pesticides, manure or silage effluent.
- Need to plant up pond with aerators that will produce oxygen underwater.
- May need to lime the water if it is too acid, in the same way as soil.
- Water can reflect light, so it can increase the growth of plantings on the north shore of a pond.
- Most ponds will slowly silt up. Every 10 years or so it may be necessary to drain the pond, dig out the silt and refill. This will provide fantastic amounts of fertile silt for using as a growing medium.

Checklist for design of pond for aquaculture

Inflow/outflow: Check there's sufficient water to keep pond full. Check spillway is sufficient for downpours, and that it does not spill over the dam wall but through undisturbed soil.

Diverse habitats: Design different depths of water. 75 cm is the most productive depth as light can penetrate to that depth, and it warms. Design different gradients of margin. A steep gradient, say inside the dam wall, will be good for controlling a band of Typha, which is invasive in shallow water but cannot grow at depth.

Growing animal feed: Think about what animals you want to feed. Protein is available through high protein plants eg Elodea (Canadian pondweed), small fish, tadpoles etc. The best systems are self feeding, but you may need to design ways of harvesting.

Crops: These can be: true aquatics, eg water lilies; marginals, eg Typha, for carbohydrate for pigs or humans; water loving plants, eg mints, blackcurrants. Think about the irrigation potential for nearby crops.

Effects of wind: Ideally design so the prevailing wind will mix the layers of water in the pond and oxygenate the water.

Fish replacement: With trout you may need to restock. Carp will breed in still waters, so if there is a predator problem such as herons, it is less expensive!

Breeding ponds: A series of small breeding ponds will increase the harvest from your pond considerably (Ref 1).

Edge: Chinampas are a Mexican system of growing vegetables beside water. You get very luxurious growth but you need a constant water level for vegetables, unless the Chinampas are floating. They are good for growing willow for charcoal making – the faster the growth, the better the charcoal.

Observation exercise
10 minutes
- Observe the difference between marginal plants on the sunny side and shade side of the water.
- Observe the different plants growing in different depths of water.

Description of the system we have at Ragmans
10 minutes with a further 5 minutes for questions
This is best done at the pond side. It can be adapted to any reasonably sized body of water, and slides are an alternative if there is no living example available.

   We use the pond at Ragmans for:
- Irrigation by gravity of a market garden, using a low pressure system from a siphon.

- Being able to put out farmyard fires with the water if necessary.
- Raising fish for consumption and for anglers.
- Growing willow for ceramic stoves and drawing charcoal.
- Growing pig and chicken feed (Typha and Elodea).
- Wildlife.
- Swimming pool.
- Soft fruit around the north side of the pond.

Links to ethics and principles

Aquaculture is a good example of stacking, cycling of nutrients, soil building, and nutrient sinks, especially phosphates. It is one of the fastest ways of creating a Zone 5 wilderness area and introducing wildlife into a system, for observation or harvest.

Living examples

Ragmans Lane Farm, Lower Lydbrook, Glos, GL17 9PA. Tel: (01594) 860244
Ruskin Mill fish farm, Nailsworth, Stroud, Glos

Further activities

If the course is taking place over a period of time the simplest thing is to measure growth in aquaculture systems. Even during a two week course there will be substantial growth. Our willow coppice grows about an inch a day during the summer. Simple measurements like this can drive home the productivity of aquaculture.

- Measure willow.
- Measure the temperature at different depths – very interesting if you can do this over a period of 10 weeks or so in spring or autumn.
- Prepare chicken or pig food from water plants: chop and mix with barley or wheat.
- Basic siphoning and water transport systems for irrigation.
- Can do basic nitrate tests (available from aquarium shops) to see if water is clean, getting cleaner or getting more polluted.

Agenda 21 link

Chapter 18(F).

Bibliography

1 Backyard Fish Farming, Paul Bryant et al, Prism Press, 1988

Further research

Farming in Ponds and Dams, Nick Romanowski, Lothian, 1994
Permaculture: A Designers' Manual, chapter 13.

Tools, Use and Care

Graham Bell

Why I include this session

It is easy to assume that just because you know how to use a tool, everyone else does. In fact we are not born with this gift, we learn it, but only if we are given an opportunity to do so. This session gives the teacher a more accurate understanding of the skills levels of their students and allows more skilled students an opportunity to share them with others.

Objective

To demonstrate appropriate choice and use of tools, including health and safety aspects.

Learning outcomes

By the end of this session students will be able to:
· State the main purpose of using tools, and principal considerations in their safe and efficient usage.
· List their existing skills, and those open to improvement.

Context

First day (see 'The First Day', pages 12-19).

Duration

35 minutes

How I teach this session

Demonstration of tools use and care, making the following points:

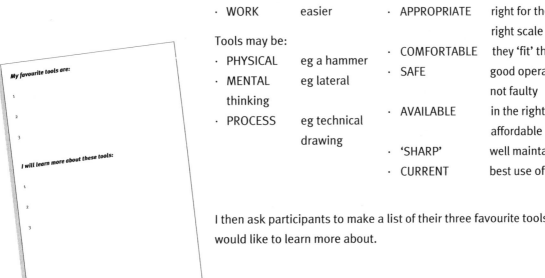

Tools make:		Tools needs to be:	
· WORK	easier	· APPROPRIATE	right for the job
			right scale
Tools may be:		· COMFORTABLE	they 'fit' the user
· PHYSICAL	eg a hammer	· SAFE	good operating knowledge
· MENTAL	eg lateral		not faulty
thinking		· AVAILABLE	in the right place
· PROCESS	eg technical		affordable
	drawing	· 'SHARP'	well maintained
		· CURRENT	best use of available technology

My favourite tools are:

1

2

3

I will learn more about these tools:

1

2

3

I then ask participants to make a list of their three favourite tools, and three they would like to learn more about.

What Do I Eat?

Patrick Whitefield

○ Objective

To get students thinking about the ecological and ethical impacts of the foods they eat.

○ Learning outcomes

By the end of this session students will be able to:
· Explain where the bulk of their food comes from and the ecological and ethical effects of their eating habits.
· Consider possibilities for changing their eating habits.
· Raise issues about the food system.

○ Context

Usually on the second day of the course, as a contrast to an information-dense session on gardening.

○ Duration

30-35 minutes

○ How I teach this session

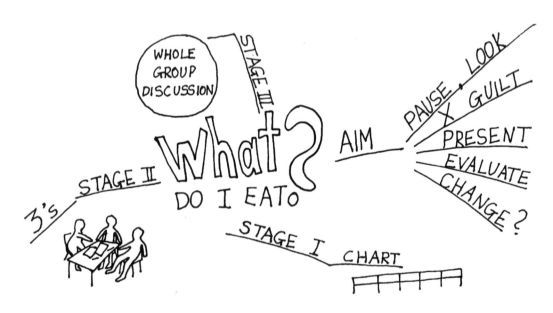

Part 1: Paper exercise

10-15 minutes

Ask the students to each take a piece of A4 paper, place it with the long axis horizontally and divide it into seven columns. Then ask them to head the columns: food; immediate source; origin; processing; organic?; how I feel; change?

I then explain column headings as follows:

In the first column write down the ingredients of the last meal you ate at home. Opposite each ingredient write down:

- Col 2, where you got them from, eg your own garden, local shop, supermarket.
- Col 3, where the food was grown.
- Col 4, processing, eg, for bread, milling and baking; you may also like to include packaging here.
- Col 5, whether it was organically grown or not.
- Col 6, how you feel about eating that food.
- Col 7, for those foods you feel less than totally happy about eating, things you might like to change.

If you finish that before the time is up, make a list of the ingredients of your second-last meal, and so on.

I always make the point that the purpose is not to make people feel guilty. With the right approach people feel empowered by this exercise.

Part 2: Small group discussion
10-15 minutes
Form groups of three and discuss findings.

Part 3: Whole group discussion
Optional 10-15 minutes
Ask if anyone has got anything arising from that discussion they would like to say to the whole group.

Agenda 21 link

Chapter 4.

Further research

Local Harvest, delicious ways to save the planet
Our Ecological Footprint
Sustain, 94 White Lion Street, London N1 NPF
Soil Association, Bristol House, 40-56 Victoria Street, Bristol, BS1 6BY. Tel: (0117) 929 0661.
Email: soilassoc@gn.apc.org Promote box schemes, local food links and keep lists of schemes in operation and planned.

17 Essential Foods

Andrew Goldring

Objective

To open up discussion around nutrition, and reinforce the thinking around bioregions. To introduce the concept of invariance.

Learning outcomes

By the end of this session students will be able to:
· List 17 essential foods for optimum health.
· Explain the concept of invariance.
· Make more informed decisions concerning nutrition.

Context

I put this in whilst covering Zone oo, and just before lunch.

Duration

45 minutes

How I teach this session

Preparation
Five days before the session start sprouting seeds, and sometime before it assemble the rest of the foods, which are: cucumber, hazelnuts, capsicums, dried olives, watercress, mung beans, kelp, alfalfa, leaf dulce, pine nuts, beets, Jerusalem artichokes, papayas, pumpkin seeds, cabbage, carrots, corn.

Exercise
10 minutes
Pass around small bowls containing the foods for the group members to taste. If you want to spice up the session get them to do it blindfolded – this guarantees laughter. Ask the group to identify each food. They can also be encouraged to mind-map them without using words.

Talk
15 minutes
I now introduce the concept of invariance and how it can be used to help us as designers. Invariants are constants or dynamic patterns which can be used by designers as checklists. The invariant in this exercise is that all human beings require 59 elements in order to stay healthy. These are provided by the foods above, which each contain all 59 elements (Ref 1). Other invariants are that we all require food, nurture, shelter etc. It is highly desirable for designers to assemble lists of invariants for personal, social and environmental ecologies, as they can be a useful guide when designing. Invariants are like islands of stability in seas of change!

Discussion

At this point the session can go a number of ways, depending on what has already been covered:

· The link between what we grow and our nutritional needs.
· Linking the list of foods to the concept of bioregions. We can't grow olives, do we trade or just eat hazelnuts?
· Other aspects of diet can be covered: eating more raw food (a PMI on this is useful), food combining and so on. Students often know loads about nutrition, so a brainstorm can be very handy here.
· Indoor gardening – sprouting seeds.

Closing

At this point the session's subject matter turns into lunch.

Notes:

1 Hybrid varieties do not always contain all 59 elements.
2 Eating two or more of these foods each day, should guarantee never becoming deficient in any element.
3 The foods mentioned above would obviously only form part of a balanced diet.

○ Agenda 21 Link

Chapter 6.

○ Bibliography

1 Diet and Salad, Walker, Dr N, Norwalk Press, 1971. Available from the Wholistic Research Company – see book and video suppliers list on pages 371-372.

○ Further research

Ageless Ageing, Kenton, L, Vermillion, 1995. Leslie Kenton researches her work brilliantly, and I recommend her many books on nutrition and health issues

'The Indoor Garden', Issue 9 & 'In The Raw': A High Energy Diet, Issue 15, Permaculture Magazine, Permanent Publications

⇨ *Local Food Links*

Matt Dunwell [many thanks to Mark Fisher]

○ *Why I include this session*

It's central to the local economy. Any discussion of how we eat, how we spend our money, and the future of agriculture should involve food links. It includes a growing number of initiatives such as box schemes, farmers' markets, local food directories, food co-ops for consumers, and producer co-ops. Although the session is tilted towards growers, it is really aimed at people who are buying food, and the choices they have. Organic, local, non-GMO, this is where the fate of the environment hangs, and it is here that the permaculture syllabus works its hardest.

○ *Objectives*

This session aims to reorientate people's buying habits towards sustainable targets.

○ *Learning outcomes*

By the end of this session students will be able to:
· Explain how a box system operates.
· Know how to encourage a box system in their area as a consumer.
· Think clearly about the impact they have on the food economy through their consumer decisions.
· List options that can encourage local food production.
· Explain why food is a community issue and how this can be used to weld a community behind a common issue.

○ *Context*

It can be good to keep it until the end of the course, because it is an excellent way to bring together a range of subjects, including: diversity, food miles, welfare (cutting down excessive transport of animals) and building community.

○ *Duration*

90 minutes

○ *How I teach this session*

Group work
20 minutes
Get people talking, in small groups, about where they get their food and why. This often brings up interesting linkages. Bring the points together with the whole group. (See also 'What Do I Eat?', pages 262-263.)

Benefits of local food links – interactive talk with slides, and/or site visit

45 minutes

A garden growing for a box scheme is much more diverse than one growing for a wholesaler or supermarket. I illustrate this by teaching this session on-site in the garden here at Ragmans Lane Farm, which produces vegetables for a box scheme. Alternatively the session could be accompanied by slides, or there could be a visit to a veg box garden at some time during the course.

Main points:

Customer

- Usually cheaper than organic prices.
- Food often delivered.
- Can go and see how food is produced rather than relying on organic symbol.
- Connection with land – ask for farm open days etc.
- Eating with the season.
- Salads and other veg often picked that day – much higher nutrition, especially vitamin C.
- Good reason for a party! – important community building tool.
- Not reinforcing monoculture and supermarkets, but gaining control yourself.

Grower:

- More profit (no middle man), making small-scale production viable.
- Less outgrades.
- More robust customer base.
- More control over what you can grow, unlike supermarkets.
- Diverse system makes it easier to control pests.
- Harder management if the system is complicated.
- Can sell other products, such as fruit and eggs.
- Favours holdings with different animals, giving more opportunity to cycle nutrients.

Farmer – the advantages of 'stacking' a 5 acre veg box scheme into a 100 acre farm include:

- Increased business to the farm gate.
- Good rent.
- Share produce delivery, machinery and labour.

Other strategies to cover:

- Farmers' markets.
- Local food directories.
- Food co-ops for consumers.
- Producer co-ops.

Buying locally

10 minutes

On the blackboard, compare the fate of '100 spent in a supermarket, with '100 spent with a locally-resourced business such as a box scheme. With a supermarket 80% of the money immediately leaves the local community, while only 20% is retained, mainly to pay wages. With a locally-resourced business the proportions are reversed. After four purchases from either kind of business the amounts of money which have circulated in the community are very different:

Supermarket				Locally-Resourced Business		
Spent	Leaves	Stays		Spent	Leaves	Stays
£100.00	£80.00	£20.00		£100.00	£20.00	£80.00
£20.00	£16.00	£4.00		£80.00	£16.00	£64.00
£4.00	£3.20	£0.80		£64.00	£12.80	£51.20
£0.80	£0.64	£0.16		£51.20	£10.24	£40.96
Total spent locally		£24.96				£236.16

Whole group discussion

15 minutes

To bring the session to a close, ask students to name the strategies we can use to increase local food links and develop the local economy.

Community Supported Agriculture, CSA

These are schemes whereby customers take a more active role in the production of their own food. In some cases they take part in deciding what crops are grown. They often pay money up front to cover the cost of growing the crop, and in return take a direct share of the crop, rather than paying for it by the pound or kilo at harvest time. This can be very good value in most years, but if a crop fails the customer shares the burden. It is a good way to capitalise and support vegetable growers who perhaps have access to land but no capital. Importantly it involves the consumer in a more direct way, and can help them to understand much more about how their food is grown.

Living examples

- Hundreds. Anything I put here will be out of date. For an up to date list contact: The Soil Association, Bristol House, 90 Victoria St, Bristol BS1 6DF. Tel: (0117) 929 0661
- Ragmans Lane Farm, Lower Lydbrook, Glos, GL17 9PA. Tel: (01594) 860244. We operate two box schemes, selling vegetables and meats.
- Devon County Council is promoting local organic food links through LA21.
- Organics Ltd, in partnership with Brighton and Hove Council, is developing a flagship LA21 project on 10 acres of land which will include a box scheme and municipal composting.
- Bradford Environmental Action Trust (BEAT) is one of the best examples of community involvement in food production in an urban or peri-urban environment. Tel: (01274) 745123.

Further activities

Design a direct selling system for an existing holding.

Selling direct requires a highly designed system, but the structures are mostly invisible. There are physical structures such as packing sheds, storage space, freezers for meat, Environmental Health Regulations etc. But more importantly there are the delivery mechanisms. Designing invisible structures is quite a challenge, but they're what makes the whole system work.

Discuss this. Work out what structures you need. How do you get the food from the field to the kitchen table? Who is delivering? Is it viable? How is it financed? What safeguards do the growers need? What standards do the consumers want? What are the feedback mechanisms? What really happens to food miles? (They may go up!) Could be lively!

○ ### *Agenda 21 link*

Chapter 18.

Food co-ops are a way of building community, and once created they can look for growers to buy direct from. Local food links are seen to be an integral part of the Agenda 21 process, and local trading systems are an ideal way to move from discussion into practical action using the Agenda 21 vehicle.

○ ### *Further research*

The Forest Food Directory, Dunwell, M, & de Selincourt, K, self-published, 1997. A directory of local food producers in the Forest of Dean. Won the National Grid Community 21 Award for best practice
Local Harvest, delicious ways to save the planet
Local Food Links, Steele, J, Soil Association, 1994
Farmers' Markets, The UK Potential, Chubb, A, co-logic books, 1998
'Welcome to the Feast of Dean', Issue 16, Permaculture Magazine, Permanent Publications, 1997. The story of the Directory and a general look at direct selling
Sustain, 94 White Lion Street, London N1 NPF

⇨ *Energy*

Patrick Whitefield

○ **Why I include this session**

Energy use is fundamental to sustainability in general and permaculture in particular. Although the subject is covered in the Principles session, another, more detailed session is necessary to cover the key concepts.

I include the case study of Jean Pain as a lead-in. This case study illustrates many other aspects of permaculture apart from energy, and is an inspiring story in itself.

○ **Objective**

To give the fundamental concepts about energy, and to get students thinking about the subject within the framework of these concepts.

○ **Learning outcomes**

By the end of the session students will be able to:
· State some of the fundamental principles of energy use.
· Approach decisions on energy use knowing what questions to ask.
· Access more detailed information on personal and household energy use.

○ **Context**

Around the middle of the first week. It doesn't depend on any other session, but should be done while the students are still fairly fresh and receptive to a solid input of information.

○ **Duration**

90 minutes

○ **How I teach this session**

I aim to get across the most important concepts. Detailed information is only given by way of examples. The Centre for Alternative Technology (CAT) do an excellent job of providing detailed information on renewable energy technology, and there's no point trying to duplicate that (Ref 1).

The concepts covered are relevant both to making decisions in our own lives and to taking an informed part in public debate.

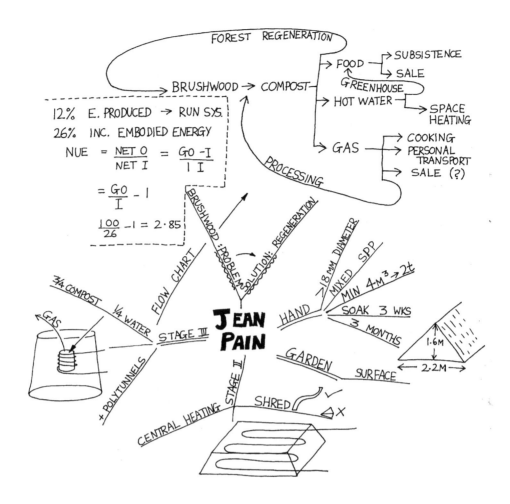

Part 1: Jean Pain, talk

25 minutes

Briefly review:

· The problem and the solution
· Hand method; emphasise key points
· Brief reference only to the gardening method
· First-stage mechanised method
· Fully integrated mechanised method

Draw a chart of the multiple outputs from this system, with cyclical movements of energy (see diagram , and Refs 2 & 3]. Emphasise the Net Useful Energy Calculation

Part 2: Jean Pain, discussion

15-20 minutes

"Get into groups of three and see if you can think of any problem in your lives or communities which could be turned into a solution."

People sometimes find this difficult. If I see any group is getting stuck I may help them, by asking some open questions (see page 17), or reassure them by telling them they're not the first group to find it difficult. Sometimes the latter is all they need to get going again. When discussion seems complete I ask if any group of three has anything to report back to the group as a whole.

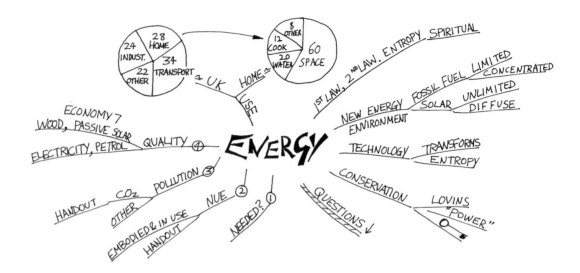

Part 3: Energy, talk

35 minutes

Hand out sheet Energy: Definitions & Figures. This reduces the students' need to take notes.

Introductory points:

· First Law of Energy (thermodynamics).
· Second Law.
· Entropy.
· Spiritual energy does not follow these laws
· The new energy environment: inverse proportion of the abundance and the concentration of fossil. fuels and solar energy (Ref 4).
· Technology a transformer not a producer.

Main point:

· Conservation: using less is the key, rather than producing more.
· Amory Lovins on the new power station (Ref 5).

Asking the right questions about an energy source:

· Is it needed?
· Net Usable Energy ratio.
· Pollution output.
· Energy quality; matching quality with use.

Energy usage:

· The national pie-chart.
· The domestic pie-chart.

Part 4: Energy, discussion

10-15 minutes

In groups of three: "Can I make use of these concepts in my own life?"

Note: this discussion may be dropped if people seem tired and ready for their cup of tea.

○ Living examples

CAT.

Brenda Vale's Autonomous House.

○ Agenda 21 link

Chapters 4(B), 7(E) & 9(B1).

○ Bibliography

1 CAT Publications. A range of titles on specific subjects, for both householders and specialists. High quality information and up-to-date lists of contacts, suppliers, sources of information, products, courses, publications and demonstration site. Contact: Centre for Alternative Technology, Machynlleth, Powys SY20 9AZ. Tel: (01654) 702400

2 Another Kind of Garden, Pain, I & J, self-published, 1972/79. Complete details of the method, including hot water and gas production

3 Comite Jean Pain, Avenue Princesse Elizabeth 18, B1030, Brussels. Tel: (00 35) 2241 0820. Promote and run courses on the Jean Pain composting method

4 Entropy, a New World View, Rifkin, J, Paladin, 1985

5 Living in the Environment. Chaps 3, 17, 18 & 21, especially p75-77 (Amory Lovins on the 'New Power Station')

○ Further research

'Energy and Permaculture', Holmgren, D, in The Permaculture Edge, Vol 3, No 3, 1993

Short Circuit. Chapter 5, review of renewable energy technologies

Home-Grown Energy from Short-Rotation Coppice, Macpherson, G, Farming Press, 1995

The Efficient Alternative, how we can do away with nuclear power, reduce pollution, conserve natural resources and save money, Agnew, PW, Tarragon Press, 1994

Factor Four, doubling wealth, halving resource use, von Weizsacker, E, Lovins, AB, & Lovins, LH, Earthscan, 1997

'Hot Water from Sunbeams: DIY Solar Water Heating System', Issue 15, Permaculture Magazine, Permanent Publications, 1997

Alternative Technology Association. Supporters' organisation for CAT

National Energy Foundation, 3 Benbow Court, Shenley Church End, Milton Keynes MK5 6JG. Tel: (01908) 501908. First stop for domestic energy advice

Energy

Peter Harper

Why I include this session

In the UK we use four or five times more fossil fuels than our sustainable 'quota'. Irreversible climate change through excess CO_2 production is potentially the worst of all environmental impacts. Therefore using less fossil fuels ranks higher environmentally than almost anything, and should have a commensurately large presence in the Design Course.

Energy is so cheap by historical standards that it is flagrantly abused. Most people have no idea of the relative power ratings or total consumption of the technologies or service systems on which they depend. The Design Course could help to spread some enlightenment here.

Basic energy literacy should be a cornerstone of permaculture, certainly for permaculture teachers. David Holmgren derived many of his original ideas from the ecological energetics of Howard Odum, but this side of permaculture has been almost entirely lost.

What's more, it's practical: students might actually save quite a bit of money on energy bills.

Objective

To ensure that design course graduates have at least a minimum level of energy literacy.

Learning outcomes

By the end of this session students will be able to:
· List the basic units of energy and power, and how energy is typically used in the home and in the wider society.
· Evaluate personal energy consumption and identify steps to reduce it.
· Rank energy alternatives in terms of their cost-effectiveness.

Context

In the middle of the course.

Duration

4-5 hours

How I teach this session

Part 1: Background information

Interactive talk
60 minutes
This is to provide basic energy literacy, and to give students the tools to make their own assessments.

- Basic calculations. Warning: when you do the numbers it often comes out differently from what popular green thought leads people to expect. This might challenge some cherished assumptions, so you need to be well prepared for arguments!
- A bit of history:
  - Past trends in energy use and what's happening now, eg energy consumption is increasing rapidly in the Third World but not in the developed world: why?
  - The energy pie charts: where it comes from and where it goes to, what share householders are directly responsible for.
  - What the various futures for energy could be and how to influence things yourself.
- The sexy topic of renewable energy, and why it's not high on the list for householders.
- Jargon, units and basic distinctions:
  - Watts and Joules.
  - Difference between energy and power, between kW and kWh.
  - Energy density, the notion of embodied energy.
  - Difference between primary, delivered and useful energy.
- Elementary calculations:
  - How to calculate total energy from power ratings.
  - Examples of how naive greenery can use more energy rather than less. Paradoxes: microwave, toaster, shower.
  - What are the big consumers in your life and which things are not worth bothering with, eg a power-mower or a shredder is going to make a negligible difference to your energy consumption, so don't insist on a hand mower on energy grounds.
  - Evaluation of various fuel sources: coal, oil, mains gas, bulk gas, bottled gas, wood; and the appliances associated with them.
- The psychology of being and feeling warm, handy tricks.

I often use simple objects to illustrate various things: hand-operated devices such as a dynamo torch, lumps of coal, or a 5 litre can of petrol. I also use overheads with tables of various measures for energy saving, what they cost and what they achieve.

Part 2: Exercises

First exercise

30 minutes

Evaluation of students' personal energy consumption.

Get them to bring gas and electricity bills, old MOT certificates, records of plane flights etc.

- A typical household breakdown. How do you compare? What are the big areas?
- What is the difference in the class between the highest and lowest users? Notice the relationship to income.
- Notice that the big things are transport and heating, although diet is one of the largest hidden energy factors, especially commercial meat.
- If there is no session in the course on transport I would strongly emphasise the role of the car in the total energy balance of a household.

On the basis of all this information a course graduate should be able to reduce energy consumption by 10-20% immediately, and by 50% within two years, and they should have the skills to monitor the process. They should log the process as part of their follow-up work.

Second exercise

30 minutes

Have the students 'spend' a globally fair and sustainable personal energy budget for the year – say 12,000 kWh or 43 GJ. It's tight, so what do you give up? You would need to have some figures of the energy cost of various activities ready to hand. A discussion could then look at the implications of various choices.

Cutting Energy Use

The big things we use energy for directly vary quite a lot from one household to another but typical figures are: transport (40%), house heating (35%), water heating (10%), appliances (10%), cooking (5%).

We're aiming eventually at reducing energy consumption to an average of 20% of present levels, although 50% is a reasonable target to start off with. If you get rid of the car and don't fly to sunny places in the summer you've got pretty close to 50% in one go. If you switch to gas, buy lots of thermal underwear and keep the house at 15°C instead of 20°C you're getting down towards 30%. If you invest in low-energy light-bulbs, eat more raw food, use showers instead of baths, insulate the loft and buy less products and more services you're probably pretty close to 20%, and it's time for you to go and sort out your water or solid waste!

Part 3: Practical measures

Interactive talk

60 minutes
· Household-scale renewables are not cost-effective unless you're living in a very remote area. Renewables do not deserve to take up more than a few minutes.
· Retrofitting, including: heating controls, thermostats, pipe and tank lagging, draughtstripping, secondary glazing, curtains, insulated shutters, targeting heat, types of insulation.
· Conservatories.
· Energy-efficient design of new houses.
· The importance of vehicle maintenance and tuning, bikes and bike maintenance, public transport.
· Thermal underwear and clothing, haybox, low-flow shower heads, off-peak electricity, sussing out household temperatures with digital thermometers – where's it all going?

Part 4: Hands-on

Practical activities

90 minutes
Learning practical skills, such as:
· Basic draughtstripping
· Thermostats and heating controls
· Insulating a hot water pipe

○ Agenda 21 link

Chapter 4(B), 7(E), 9(B1).

○ Further research

Energy – A Guidebook, Ramage, J, OUP, 1997
Save Energy, Save Money, Jackson, F, CAT Publications. A practical guide to energy in the home, which also contains a comprehensive resource guide to organisations, products, publications and courses
CAT also has special resource guides on Transport and Sustainable Activities
Renewable Energy, ed Boyle, G, OUP, 1996

Most regions have an Energy Advice Centre, see Yellow Pages or list of addresses in Save Energy, Save Money.

There are no short courses on energy literacy, not even at CAT. A good reason to include it in the Design Course? If you are particularly interested in renewable energy, CAT is the place, with dozens of courses and publications.

My Dream Home

Graham Bell

○ Why I include this session

Because it's liberating and empowering.

○ Objective

To demonstrate to people their ability to see their dreams as possible.

○ Learning outcomes

By the end of this session, students will be able to:
· Understand the value of freehand drawing in liberating creative thinking.
· Know how to move beyond the obvious, to what they or a client really wants.

○ Context

Before a session on buildings.

○ Duration

20 minutes

○ How I teach this session

I invite students to draw their Dream Home on a very big (AO) sheet of paper with kid's crayons.

Using very big paper and large kid's crayons is important, as the aim of the session is not that they end up with technically accurate renderings of a house they would like to live in, but rather that they open up their imagination and respond to deeper feelings and needs. By identifying these deeper needs before we go into the 'technical' building session, participants can take their sense of vision with them. This opens up creative possibilities.

To draw the exercise to a close, we share observations about the drawings and what they may be telling us. This is done by each person explaining their own drawing, and others being asked if there is anything they particularly admire about the presentation. There is no criticism. You can't ask people to share their (vulnerable) dreams and then tell them they're wrong!

Many students don't draw buildings, but place themselves under trees, and a large number of the houses that are drawn are round and often womb-like. This is revolutionary compared to the reality of most buildings. But whatever comes out, just accept it.

This exercise concentrates people's understanding of 'why build', rather than just 'because we've always done it that way.' The by-product is to reinforce people's sense that they can achieve what they want.

Buildings and Structures

Simon Pratt

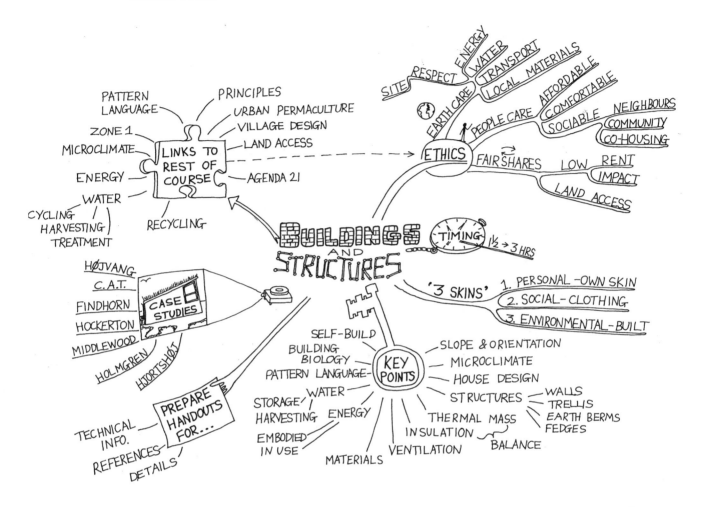

Why I include this session

Shelter is one of the fundamental human needs. Everyone lives somewhere and the home may be a more practical starting point than land-based activity.

Objective

To explain how permaculture principles apply to where and how we live and indicate some solutions for sustainable buildings.

Learning outcomes

By the end of this session students will be able to:

· Make improvements to their own homes.

· Have access to a range of sustainable solutions.

· Explain the critical factors about energy consumption in the home and how to reduce it.

· Explain the factors involved in choosing an ideal house site.

Context

Second half of the course.

○   **Duration**

Two 45 minute sessions in the classroom, plus option of practical work

○   **How I teach these sessions**

My approach with this topic is the same as with any other. I attempt to stimulate thinking and acting in different ways, and provide some solid information backed up by references, handouts, slides, video etc. I suggest things participants can use in their lives, introducing some wacky ideas. I use case studies to make it real. How to present and what to cover will vary with the group. I try to touch on most of the ideas below, but only going into detail with one or two. I feel points 1-10 are essential, 11-14 desirable.

Key points

I emphasise the difference between new build and retrofit. The former can be easier to provide a very 'green' building, but the latter is where most of us live, and where much can be done.

1 Slope and orientation. (House sites may be covered earlier in course.)

2 Micro-climate
· Effect on house.
· Effect of house (Ref 1, p41).

3 House design
· Optimise solar gain.
· Greenhouse/conservatory on south side.
· Greenhouse as example of multi-functionality.
· Retrofitting recommended, eg window greenhouse helps where space is limited.
· Cool shade house on North side.
· Placement of rooms (Ref 1 p.71 & Ref 2, p87).
· Pond on south side will reflect light and heat into house in winter.
· Feng Shui for selection of house site and detailed layout of rooms and furniture.

4 Structures
· Increasing growing space in Zone 1.
· Providing transition from built environment to garden.
· Walls, trellis (Ref 1, p94), fedges, earth berms (Ref 1, p41).
· Living structures, especially willow, for shade, play and social space, fences.

5 Thermal mass
· Modify extremes of climate (Ref 1, p72 & Ref 4, p70).

6 Insulation
· Comparison of different materials (Ref 5, p107 & Ref 6 p206).
· Analysis of effectiveness of different insulation methods (Ref 7, p10).
· Ground under solid floor as heat store.

7 Ventilation
· Super-insulated buildings.
· Condensation – ventilation will reduce problems (Ref 8, p48).
· Passive.
· Mechanical.

8 Materials

- Examples of vernacular buildings. Encourage participants to think about their own locality (Ref 2, p84).
- Reuse and durability of buildings – what are the environmental impacts when it is demolished? (Ref 9, p42).
- Discussion of different materials (Ref 5, p91 & Ref 8, Part 4).

9 Energy in buildings

- Distinction between embodied energy of construction materials and energy used for heating and appliances.
- Ultimate low energy building uses materials collected from site.
- Comparison of energy content of different materials (Ref 5, p92; Ref 9, p41 & Ref 6, p205).
- Energy conservation priority over production.
- Heat type, radiant rather than convected heat.
- Draught-proofing and insulation.
- Lobbies at entrances.
- Curtains, shutters and blinds.
- Secondary glazing.
- Low energy appliances, including light bulbs.
- Passive and active solar heating systems, photovoltaic cells (Ref 8, Part 2).

10 Water (May be covered earlier in course)

- Collection of rainwater from roof.
- One third of household water is used for flush toilets.
- Grey water can be used directly in the garden or in toilet.
- Separation of potable water from other supplies, eg washing, garden.
- Use of filters.
- Discussion of ways to reduce consumption, eg Robert and Brenda Vale's autonomous rainwater. collection system.

11 Pattern language

- Work of Christopher Alexander et al.
- Present an example to stimulate discussion; eg 106 Positive Outdoor Space – outdoor spaces which are merely left over between buildings will, in general, not be used (Ref 3).

12 Building biology (Baubiologie)

- Likens built environment to our 'third skin' (first is our own skin, second our clothing).
- Buildings need to be able to breathe and function as regulator, protector and insulator.
- Breathing wall developed at Findhorn (Ref 6, p52).
- Healthy living environment, geopathic stress and electromagnetic fields.
- Sick building syndrome.
- Design for health of building and occupants. Institute of Building Biology has developed an ecological assessment of building materials (Ref 6, p20; Ref 4, p52 & Ref 11).

13 Self-build

- Historical context.
- Walter Segal method (Ref 10).

14 Miscellaneous

- Temporary shelters.
- Underground buildings.
- Earthships.

Further activities

· Design layout of kitchen for optimum use.
· Locate ideal house sites on map, 3-D model or on the ground.
· Build a shelter from found materials and sleep in it.
· Design and build space to use on course. Straw bale building is particularly good as it is cheap and easy. Several courses have built a structure to be used during the course.

Link to the ethics and principles

· Care of the earth: respect for the site, self-provide for energy and water, use local materials and minimise transport both during construction and in use.
· Care of people: affordable, comfortable, social – shared facilities, co-operating with neighbours, communities, co-housing.
· Fair shares: low rent, low impact, land access.

Living examples

1 Andels Society eco-village, Hjortshj, Denmark. Most houses built with rammed earth.

2 Jan and Susanne Hjvang, Skive, Denmark. 'Zero energy' family home self-providing for water supply and treatment, energy and electricity.

3 Hockerton Housing Project. A terrace of five earth sheltered houses with water collection and treatment, no external heat source.

4 Centre for Alternative Technology (CAT), Machynlleth, Wales. Several ecological buildings including a low energy Walter Segal house (Ref 12).

5 Findhorn Foundation, Forres, Scotland. The ecological houses described in Ref 6.

6 Tony Wrench, Newcastle Emlyn, Wales. Many innovative low-cost projects. The Permaculture Plot, Permaculture Magazine.

7 Rod Everett, Middle Wood Centre, Lancaster. An ecological study centre built in 1992, and a new straw bale building under construction. The Permaculture Plot.

8 Hepburn Permaculture Gardens. The ecological family home described in the book of the same name and in the Global Gardener video (Ref 13).

Agenda 21 Link

Chapter 7.

Bibliography

1 Introduction to Permaculture. Especially Chapter 3. This is more useful than the Designers' Manual – there is more material on buildings and structures in the Introductio

2 The Permaculture Way. Especially Chapter 7

3 A Pattern Language

4 The Natural House Book, David Pearson, Gaia Books, 1989. Inspiring examples with colour photos. Comparison of natural house with dangerous house, discussion of life systems and materials, design of spaces

5 Out of the Woods, Borer, P & Harris, C, CAT, 1994. Ecological design for timber frame self-build

housing using the Walter Segal method. Useful green guide to building materials

6 Simply Build Green, Talbott, J, Findhorn Press, 1995. Technical guide to the ecological houses at Findhorn

7 Save Energy Save Money, Jackson, F, CAT, 1995

8 Eco-Renovation, Harland, E, Green Books, 1993

9 Green Architecture, Vale, R, & Vale, B, Thames and Hudson, 1991

10 Walter Segal Self Build Trust, 57 Chalton Street, London, NW1 1HU. Tel: 020 7388 9582

11 Institute of Building Biology, Rectory lane, Ashdown, Saffron Walden, Essex, CB10 2HN. Tel: (01799) 584727

12 Centre for Alternative Technology (CAT), Machynlleth, Powys, SY209AZ. Tel: (01654) 702400.

13 Hepburn Permaculture Gardens

Further research

The Greener Homes Guide, Greenpeace, 1997. Excellent consumer leaflet produced to accompany the Greenpeace Eco-home, worth obtaining as a handout

Association for Environment Conscious Building (AECB), Nant-y-Garreg, Saron, Llandysul, Carmarthenshire, SA44 5EJ. Tel: (01559) 370908. Produce an invaluable directory of members (architects, builders and suppliers), Building for a Future magazine and the Greener Building Directory.

Ecological Design Association, The British School, Slad Road, Stroud, GL5 1QW. Tel: (01453) 765575. Promotes environmentally friendly design and publishes Eco Design magazine

SALVO, Ford Woodhouse, Berwick-upon-Tweed, TD15 2QF. Tel: (01668) 216494. Produce a directory listing suppliers of reclaimed materials

Zone 0: Retrofit Exercise

Bryn Thomas

Why I include this session

Bill Mollison says that in temperate climates permaculture is 50% about buildings. I could not agree more with this statement (Ref 1). Much of our negative impact on the environment is directly related to buildings and housing. Many other issues like energy, transport and food are further affected by our use and design of buildings. Furthermore our comfort, contentment and spiritual well-being are affected by our living environment, resulting in impact on the full sphere of human existence. I could not disagree more with the attitude I occasionally encounter that permaculture is about gardening and farming, and that the other bits are peripheral. The built environment is central to permaculture and sustainability.

More specifically I find that the need to retrofit existing buildings requires stressing. Many of the people on permaculture courses interested in buildings are specifically interested in new build. Whilst a new building may use less energy it will take very many years to recoup the energy embodied in its structure. It can be far greener to reuse existing buildings. Reduce, reuse, recycle, then new!

Objective

To give students an appreciation of issues relating to the reduction of the negative impact of buildings on the environment and an empowering experience of group work.

Learning outcomes

By the end of this session students will be able to:
· Show an understanding of sustainability issues relating to buildings.
· Demonstrate knowledge of practical strategies to reduce the negative impact of buildings on the environment.
· Implement some of these strategies.
· Identify relevant social issues.
· Appreciate the collective resource of knowledge in the group and the value of working and networking with others.
· Work as part of a design team containing others with complementary skills.

Context

I often include this session half way through a course. I prefer to start with ethics, principles and design. This is followed by themed days on applications, starting with soils, water, gardening etc, followed by buildings and leading into more social orientated stuff, economy, community etc. Then comes a design activity.

I like to give issues around buildings a full day, which inevitably leads into many social issues relating to building use and where we live. (See 'Example Course Format' on page 61.)

Duration

105-120 minutes

○   **How I teach this session**

Introduction to session

5 minutes

I stress the importance of retrofitting existing buildings as opposed to building new ones.

Design brief and divide into groups

5 minutes

I set them the activity of coming up with design ideas to retrofit a building. This is often the one we are using, if not one they can visit in the immediate vicinity. It is ideal to look at it being a home, but residential or day visitor centres have also worked well. Either way the building shouldn't be too large or complex.

Briefing should include emphasis on the need to reduce water and energy use, as well as using appropriate locally available sources. It may also be appropriate to stipulate that a small extension is required. This gives them the opportunity to think about conservatories.

I then divide the group into design groups of three or four. Care should be taken to place people with similar interests or backgrounds in different groups. It is possible that people who feel they don't know much could feel intimidated if they are not supported by others in the group with more confidence. Similarly if a group contains a couple of people with a hi-tech approach it is possible for them to miss many of the important points. Remember the activity is about coming up with ideas, not finished designs.

Design activity

30 minutes minimum

During the actual design activity I am active in going round the groups to give them pointers and to keep them on track and on time. It may be that they get too involved in a specific issue and don't get round to addressing, say, energy issues.

A tea break can be scheduled following this. Some groups may design through the break; others might not. I would neither encourage nor discourage this.

Report-back from design groups

30-40 minutes

Each design group can be given 5 to 10 minutes, depending on the number of design groups, to present their ideas to the group as a whole. Sometimes it is surprising how similar the different designs are and sometimes it is surprising how different they are. After each design group has presented their ideas I like to say one thing I liked about their ideas and maybe one thing I would have done differently. This gives the opportunity to stress important points. If there is time other comments and discussion from the group as a whole can be invited. However this can eat time and care must be taken to ensure time isn't lost and that comments are presented positively.

Structured summary

20 minutes

I conclude with some structured teaching. This follows topics such as micro-climate, insulation, passive solar gain, thermal mass, rainwater harvesting and working from home, using illustrations from the design. Most if not all of the issues should have been covered already, but this allows them to be considered more systematically and in a structured way for participants to take notes if they wish.

This can be followed by structured or unstructured discussion if time permits.

Links to ethics and principles

I find that the application of the principles in forest gardening comes to people immediately. In buildings their application is equally appropriate, if less obvious.

The importance of the following can all be stressed:
· Biological resources.
· Self regulating systems.
· Minimum input.
· Use of patterns.
· Diversity.
· Relative location.
· Edge.
· Energy efficient planning.
· Energy cycling.
· Low maintenance systems.
· Multiple function.

Agenda 21 link

Chapter 7.
It is important to stress that whilst this session deals with the practicalities of retrofit, the realities of achieving this are often very socially orientated and must be community led. Many people live in rented accommodation or in buildings with shared resources. This of course presents opportunities, but requires community co-operation and the right political environment. This provides a good link to future topics, including: community organisations, housing co-ops and working with groups.

Bibliography

1 Global Gardener video

Water and Sewage Systems in the House and Garden

Peter Harper

Why I include this session

- On a world scale, water is one of the big sustainability problems, and will probably be the factor that most limits food production in the coming century.
- There are impacts in getting it, and more in cleaning it.
- Conventional measures should be able to halve water consumption.
- Permaculture measures, which include lifestyle changes as well, should achieve an 80% reduction or more.
- Nutrients in waste water are a wasted resource and can cause pollution.

Objective

To introduce the water cycle as it applies to buildings and to examine improvements in efficiency and alternative possibilities for both supply and treatment.

Learning outcomes

By the end of this session students will be able to:
- Give general statistics of water use.
- List basic methods for reducing consumption.
- List basic principles of rainwater collection and greywater re-use.
- Explain the basic systems for dealing with waste water and sewage and in what circumstances they are appropriate.

Context

Day 4 or 5.

Duration

Two 60 minute sessions

How I teach these sessions

Part 1: Water supply

Information provision with some discussion

60 minutes

Graphic pie-charts of domestic, national and international water use would be very useful.
- Global water availability:
  - where it comes from.
  - how it's used, industrially, agriculturally, domestically etc.
  - who gets how much.

- Water use in UK households:
  - how it can vary a lot but there are some pretty solid patterns.
  - how it's charged for.
- How to reduce water usage:
  - simple habits.
  - low-flow appliances.
  - basic cost-effectiveness calculations for various measures.
  - getting a meter installed and monitoring your own use pattern.
- Collecting water yourself:
  - rain, wells, springs, streams.
  - the legal niceties and what you need to know.
  - getting the water tested.
  - purification measures for drinking.
- Special attention to rainwater for irrigation and a few other uses in the house:
  - what you could expect to get.
  - collecting, filtering and storing it.
  - levels, pressure, pumps.
  - basic calculations.
- A sober view of water alternatives: is it worth the cost and hassle?

Part 2: Dirty water

Information provision with some discussion
60 minutes

- Waste water:
  - the different sorts, the likely proportions of each, what's in them.
  - evaluating what's worth treating or recovering.
- Minimising waste water:
  - low-flush and urine-separating toilets.
  - dry toilets 57 varieties, pros and cons.
- The nutrient content of grey water, urine, faeces and other biological wastes in houses and gardens.

Nutrient recovery

Ninety-five per cent of the nutrient output from a house and garden are in urine and the easiest-to-collect fractions of kitchen and garden waste. Generally there's enough N, P, K, Mg, Ca and countless trace elements in a family's urine to fertilise any amount of garden. If you collect urine systematically you end up with colossal amounts and it's quite a problem to work out what to do with it. We are working on it at CAT but have only provisional results to report. A terrific area for permaculture research.

Trying to recycle faeces and grey water is not usually worth the effort – a mackerel to catch a sprat.

- Grey water:
  - various things that can be done, mostly to benefit the garden.
  - the need to keep it simple.
  - still a lot of unknowns, and techniques in a process of trial and development –
    a good area for research.

· Urine in the garden
  · again a lot of unknowns and a good area for research.
· If you are not on the mains sewage:
  · if you have a dry toilet you still have to deal with other waste water to the satisfaction of the Environmental Health and the Environment Agency.
  · ponds, reedbeds, septic tanks, leachfields etc: space required and approximate costs.

Part 3: Fringy stuff

15 minutes if there's time
· The deeper structure of water, flow-forms, homeopathy, Grander water, Plocher water.

○ Living examples

Centre for Alternative Technology (CAT), Machynlleth, Powys, SY20 9AZ. Tel: (01855) 831655
Elemental Solutions, Oaklands Park, Newnham Glos GL4 1EF. Tel: (01594) 516063
The Water Association, c/o Ruskin Mill, Nailsworth, Glos
Ragman's Lane Farm, Lower Lydbrook, Glos, GL17 9PA. Tel: (01594) 860244

○ Further activities

At least one practical activity should be included with these sessions. Possibilities include:
· Rigging up a greywater diversion system.
· Making a simple greywater straw filter.
· Measuring runoff from a roof.
· Making a simple peecan.
· Fitting low-flow shower and tap heads.
· Unblocking a drain with rods, the plunger.
· Elementary plumbing with plastic waste-pipe.

○ Agenda 21 link

Chapters 7(D), 18(D), 21.

○ Further research

When I teach these topics at CAT I have the benefit of a great variety of real systems and specially-constructed models to demonstrate them. I also have lecture-notes, overheads and slides, which I could make available to other teachers. I would recommend prospective teachers attend suitable courses on the subject and read the basic books.

Cheap CAT tipsheets:
Water Conservation in the Home
Come Clean (washing machines)
Making Use of Waste Water
Slow Gravity Sand Filters
Constructed Reed Beds

CAT Resource Guides with sources and references:
Water Supply and Treatment
Sanitation

Safe to Drink, Stauffer , J, CAT Publications, 1996. Water quality
Sewage Solutions, Moodie, M, Grant, N & Weedon, C, CAT Publications, 1996. Reed beds and other alternatives
Fertile Waste, Harper, P & Thorpe, D, CAT Publications, 1994. Dry toilets
Create an Oasis with Greywater, Ludwig, A, Oasis Design, 1997. Best treatment of greywater use

Courses on water supply and on sewage alternatives are held at CAT and Elemental Solutions.
For esoteric aspects of water, flowforms etc, contact The Water Association (Address under 'Living Examples', above)

Recycling and Solid Waste

Peter Harper

○ Why I include this session

It's fairly easy to get the output of solid waste from a household down to 20% of average or even lower, so let's do it. But don't do it beyond the point where it becomes a hassle. Probably in terms of its overall environmental impact solid waste ranks fairly low, so it's worth basic good practice, but not losing a lot of sleep over.

Chucking things away feels bad, and many people in our movement are inclined to hoard materials that "may come in handy one day", or to make excessive efforts to reuse or recycle stuff. By excessive I mean efforts which incur greater environmental costs than just chucking them away would; or which take time that could be used for greater environmental benefit in other directions, or which just clutter up your life. We must not attempt to reduce waste to zero. Down to 20% is fine to be getting on with.

This is an example of my Quit While You're Ahead principle. (See 'What? Car Maintenance in a Permaculture Design Course!' on pages 36-42 for an explanation of this principle.)

○ Objective

To introduce the basic forms of household wastes and the most sustainable ways of dealing with them.

Learning outcomes

By the end of this session students will be able to:

· Sort mixed household, garden or construction waste into reusables, recyclables, compostables and refuse.
· Assess when further effort in waste reduction is not justified.

○ Context

Middle of the course.

○ Duration

60 minutes

○ How I teach this session

Interactive talk

60 minutes

I use pie charts throughout to establish the proportions. I make a lot of use of my own experience with the solid waste from a house conversion and two years of closely monitored house and garden waste. Here are the main points that I make:

Getting a perspective

· Household solid waste as a proportion of total solid waste. Although it's a relatively small part of the total, it's often quite nasty, has immense symbolism, and we have direct control over it.

Indoor waste

· What the typical dustbin contains:

- how much could be avoided by attention to the quantity and quality of what we buy.
- what can potentially be reused, recycled or composted.
- reusing can be fun (bottle corks for insulation, plastic bottles for cloches etc) but as soon as it gets to be the slightest worry or hassle it's counter-productive.
- composting is the best solution for all organic material.
- Establish what happens to your rubbish after collection, and what recycling facilities there are – phone the council. (See box, 'Rubbish Realities'.)
- The importance of family culture: working waste management around things that people, especially kids, like to do – crushing cans, putting things in recycling banks, squashing and buckling cardboard boxes.

Rubbish Realities

In most places in the UK solid waste is landfilled, in others it's incinerated or composted. Knowing what happens to your waste might make a difference to your waste strategy.

You also need to establish the distance to your local recycling facilities. In general the economics and ecologics of recycling are so finely balanced that it's rarely worth a special trip. In most areas there are recycling points for glass, paper and drinks cans not far away. There may also be steel cans and plastic bottles. Take a wheelbarrow.

If you have a garden, most of the rest can be composted. If you have no garden and there is no civic amenity site or community compost scheme, try to find a compost enthusiast who will take your kitchen waste, or offer to run their compost system for them. Note that nearly all cardboard and soiled or crumpled paper can be composted as well as the usual food scraps. People will ask about the inks. Everybody thinks they are deadly but they are not these days (Ref 1.).

If you've done all the above most of your waste will now be plastics, generally packaging. Take steps to minimise the amount of plastic coming into the house, but don't get paranoid about plastics. They have their place in modern life, and are sometimes the right choice for a particular function. Also they are relatively well-behaved in landfills: they compact a lot, are inert and take a certain amount of carbon with them that won't be going back into the atmosphere. Paradoxically being non-biodegradable is an advantage!

Basically the dustman gets about 20% by weight of my household garbage, about 8% if I include in the calculation garden waste that I might have given him if I were completely bonkers. But this 'complete rubbish' fraction is mostly inert material, worth nothing for recycling and with negligible environmental impact in the ground. So the environmental impact of my solid waste has probably been reduced to something like 3% of its potential level, with actually rather little effort.

Garden waste
- General points:
  - usually greater in volume than household waste.
  - virtually all can be composted at home.
  - woody waste can go to a civic amenity site, or green waste collection service for composting, be shredded, or used in a 'magic mound' (Ref 2).
- Composting techniques.

Building waste

Considering that many students on a Design Course would be planning eco-retrofits it's worth giving a bit of guidance about building wastes and what can be done with them. I show slides of the house conversion, how I avoided the use of skips, and reused 95% of the construction 'waste' in the garden.

· Work out a strategy first, and guidelines to follow.
· If you are not doing the work yourself make sure you let the builders know your plan and stick to it.
· The garden design may be influenced by what's coming out of the house and when, and the schedule of works in the house and garden may need dovetailing.
· 'No skips' is an excellent mind-concentrator.
· Separation of materials can be expensive both in space and time.
· Examples of sorting and reusing:
  · wood for fuel and coarse lumber (compost bins, seats, tree houses).
  · rubble for paths, wall fill, re-profiling the garden.
  · chimney pots for rhubarb or chicory blanching.
  · slates for bed edging.
  · carpets for mulches or root barriers.
  · bricks for paths, retaining walls, steps, seats, bed edges, raised beds.

○ Further activities

· During the course the waste generated is treated in an exemplary way, sorted into various categories. The one unusual area would be composting paper and card, with which I am slightly obsessed. It's actually much simpler than classical composting and ought to be routine in any permaculture household.
· The design of kitchens for recycling can be considered.
· Making a magic mound is fun.

○ Agenda 21 link

Chapters 4, 7(D) & 21.

○ Bibliography

1 Composting Secrets, available from CAT. For composting, especially using paper and card
2 The Natural Garden Book, Harper, P, Gaia Books 1994. For the magic mound technique

○ Further research

Reuse, Repair, Recycle, McHarry, J, Gaia Books, 1993. A good general book

For a blow-by-blow account of waste reduction, Complete Rubbish at No 24 is a report of my first two years of waste and what I did with it, available from CAT

Salvo. For building wastes see this amazing architectural salvage magazine

CAT's Recycling Resource Guide has lots of addresses, classified under various waste categories

Councils always give out leaflets on what you can or cannot recycle.

'Recycling & Beyond', Issue 10, Permaculture Magazine, Permanent Publications, 1995

Composting for All, starring Nicky Scott, Iota Pictures, 1997 (video)

⇨ Transport

Judith Hanna

○ Why I include this session

Car and road freight dependence have been called the worst environmental problem in the developed world. They are responsible for over a quarter of fossil fuel use, air and water pollution, road deaths (animal as well as people), land-take and severance, noise and stress.

Transport is the way a site makes links to the rest of the world. Designing sustainable transport links is an essential element in designing for sustainability. As soon as anyone steps out of their own front or back gate, they are doing 'transport'.

Design for sustainable access

Choosing a new site/location

- Choose site to minimise distances travelled
- Choose site for good sustainable access and links

Go to 'existing site/location'

Design solutions – 'green transport plan'

- Street environment: shortcuts, safe crossings, traffic calming, sight lines Community involvement in planning
- Public transport: missing services, siting of stops, signing and information (other languages?)
- Opportunities: promoting green space/links, art/imagination
- Your project: cycle parking, showers, lockers, discount admission, vouchers, refreshment for non-car visitors, promotional literature, car-sharing, delivery services, tree planting to absorb CO_2 emissions

Existing site/location

Access audit

Who needs to reach the site?
- residents (human, wildlife)
- workers   · suppliers
- customers   · visitors

Where are they coming from and going to?
- origin – destination mapping/desire lines

How often and when?
- scheduling (peak spreading/flexitime)

How to travel? (hierarchy and access modes)
- Walking (and wheeled pedestrians – wheelchairs, barrows, trolleys)
- Cycling
- Public and community transport
- Essential freight and services
- Private cars/motor vehicles

Identify barriers/missing links (in space, time and information)

Act on audit/survey findings

Review/evaluate 3-5 yearly or as required

○ Objective

To explain the need to consider transport and access as part of site design and the basic design options.

○ Learning outcomes

By the end of this session students will be able to:
· Think through the sustainability implications of their own travel behaviour and choices.
· Identify factors important to adopting 'benign' travel habits.
· Explain the main issues for designing sustainable transport/access.
· Link action to secure sustainable access with the Participative Community Planning and Official Systems sessions.

○ Context

This topic can follow on from Urban and Rural Permaculture and Local Agenda 21. It involves applying Participative Community Planning techniques and Working with Official Systems.

○ Duration

Parts 1 and 2: 30-40 minutes
Part 3: 120 minutes to half a day would serve as a practical introduction for Participative Community Planning
Part 4: 20-45 minutes

○ How I teach this session

My main teaching approach is individual and group exercises and discussion to draw out students' own experiences and insights. If course participants are involved in a shared project, the exercises 1, 2 and 4 below can focus on that, and be done in pairs or threes. If not, students get best value from focusing on a site or journey significant to them.

The Permaculture Association's Sustainable Transport Solutions briefing sheet can be used as a supplementary handout. Other organisations listed at the end also produce useful factsheets and briefings.

1 How does everyone get to this class?
· What mode of transport? How far?
· If by car: what would need to change for them to switch the journey to a more sustainable mode? What action? By who?
· If by foot, cycle or public transport: what problems did/do they encounter? What action would make using that mode easier and more attractive? By who?

Notes:
· This can be a quick starter, or can be developed into doing a mini access audit.
· Charting can include adding up miles, and total and average travel time, by each mode, and calculating the course's car fuel 'footprint'. How many trees should the course plant to cover it?
· A rough origin-destination map can be drawn, showing where people travel from.
· The 'action needed' discussion can be in small groups, one each for car users, cyclists, walkers and public transport users.
· There can be potential in a second stage, getting people into groups by the mode they'd like to use.

2 Observe and research: identifying transport problems and solutions
· Ask everyone to think of a site or journey important in their lives. What transport problems affect it? What sort of solutions could help? 2-5 minutes for them to write or draw. Brief feedback to group.

· Mind-map: types of problem identified and types of solution needed for each.

Notes:

· The individual thinking/writing/drawing element of this can be set as a homework design exercise, either as a lead-in, or follow-up to teaching this topic.

3 Participative planning for transport solutions

Transport space is shared public space, so transport solutions need to win community support. The three essential elements in a transport solution are:

i) Site design, and spatial links to other elements of the systems it is part of, eg visitors, staff, suppliers, customers.

ii) Official planning permission and infrastructure provision, eg cycle or pedestrian facilities, bus stop, schedule or route. Getting official co-operation is likely to mean running a campaign. (See 'Official Systems' on pages 340-342). Winning a campaign often depends on:

iii) Wider local support and involvement. This is most effectively generated through Participative Community Planning techniques.

Notes:

· The methods outlined in Part 2 of the 'Participative Community Planning' session
   (see pages 124-126) can be applied to designing transport solutions.

· Transport problems and solutions almost always emerge in a Planning for Real exercise, whatever its starting point.

Spending a two hour or half day session working through one or a range of Community Planning techniques is the best way to develop insight into the process of developing practical transport solutions. Particularly if the class has a shared local base or shared project.

4 Homework or individual design activity

· Develop sustainable transport solutions for own activities or project.

Campaign dates for sustainable transport promotion (for press publicity, LA21 focus, etc):
· Local Transport Day: first Saturday in March, focuses on what local authorities are or should be doing on transport. Contact Transport 2000.
· Green Transport Week and National Car-Free Day, National Bike Week, Family Rambling Day and Don't Choke Britain: cover the first to third week of June. Information from ETA, Cycle Touring Campaign, Ramblers Association and Local Government Association respectively. (See organisations below.)

○ Link to principles

· Relative location. The more that elements of a system are local or on-site, the less the need for motorised transport, and the greater scope for using and encouraging the benign modes of walking, cycling and public transport.

· Shortening and closing the cycles. What comes into the site? What goes out? How far do they travel, and how? What impacts, beneficial or damaging, are created?

· Maximum output for minimum input. Plan for proximity and accessibility. Minimise intervening distance and fossil fuel use. Minimise road and parking space, releasing land for sustainable uses. Time budget: minimise travel time, or get double use from it.

○   **Agenda 21 link**

Chapter 6 (D,E), 7 (E) & 9 (B/b).

○   **Further research**

Anyone seriously interested in pursuing transport issues in the UK should subscribe to or arrange to see the fortnightly Local Transport Today, for in-depth coverage of policy and good practice news. (£60/yr, Quadrant House, 250 Kennington Lane, London SE11 5RD. Tel: 020 7582 6626

Traffic and Transport: A Consumers Guide to Influencing the Transport System, Planning Aid for London/T2000, 1994

The Greening of Urban Transport, Tolley, R (Ed), Wiley, 1997

Travel Sickness: ...Sustainable Transport Policy for Britain, Roberts (Ed) et al, Lawrence & Wishart, 1992. Foreword by John Prescott MP

Transport 2000, Walkden House, 10 Melton St, London NW1 2EJ. Tel: 0171 388 8386; Fax: 0171 388 2481. Environmental transport campaigning umbrella organisation, hosts ALARM-UK, and 'Streets for People'

Sustrans, 35 King Street, Bristol, BS1 4DZ. Tel: (0117) 926 8893

Environmental Transport Association (ETA), 10 Church Street, Weybridge, KT13 8RS. Tel: (01932) 828882

Local branches of FOE, CPRE, Civic Trust and other broad environmental organisations will have activists working on transport issues. The national offices produce a range of useful briefing material.

The Urban Environment

Thomas Remiarz

Why I include this session

Because most students are likely to live in urban environments. Additionally towns and cities can be daunting and disempowering places, so it is important to give students confidence that they can do something to improve them.

Objective

To explore the physical and social characteristics of the city, and strategies to enhance them.

Learning outcomes

By the end of this session students will be able to:
· Think about different aspects of cities creatively.
· Apply the zoning concept in their own lives.
· Recognise how different activities can be arranged relative to each other.
· Identify resources needed to implement change.

Context

Towards the end of the course, so that design tools and principles can be applied. If Planning for Real and other participatory techniques have not been covered previously, then this session could be lengthened to incorporate them.

Duration

90 minutes

How I teach this session

Short introduction
3 minutes
Cities are a major problem and therefore need some major solutions!

PMI on towns and cities (see pages 117-121)
20 minutes
I remind students that we are thinking about both the physical and social landscape.

I go round the points raised and ask participants to develop strategies that either enhance existing Plus points, eliminate Minus points (or better still, turn them into Plus points) and further develop the Interesting points.

This is done quite quickly as a whole group.

Cities or neighbourhoods?
20 minutes
Ask the group: "How much of the city do you use"?
· daily?
· weekly?

· monthly or less?

· not at all?

Give them five minutes to complete this exercise. If students are all from the same town or city then this exercise can be done by literally mapping movements on photocopied maps, otherwise students can make conceptual sketches.

Then get them into pairs to discuss their findings for three minutes.

Collate the results by asking the whole group: "What did you find out?"

This generally results in people realising that they spend most of their time in their neighbourhood. I then make reference to zoning.

Short break

Exercise

25 minutes

I write on a board, or on cards that can be given out, the following key words: work, rest, play, learning, travel.

Now I split students into groups, one for each key word. I ask them to spend 10 minutes developing two or three proposals that could enhance their activity within the city.

Each group then reports back with their proposals.

As a whole group, make connections between these activities, and the proposals for improvement.

Next steps

20 minutes

Ask students to identify some first steps towards enhancing their neighbourhood. What can be done easily? With who? If available handouts can be distributed with useful organisations and book references.

○ Living examples

'In the City', Issue 3

'A Special Harvest: A Community Forest Garden in the Heart of London', Issue 7

'EcoEstates', Issue 8

'Incredible Edible Cities', Issue 10

'Beyond Our Own Backyard: Contagious Urban Oases', Issue 16

All in Permaculture Magazine, Permanent Publications.

Also, Bradford Environmental Action Trust (BEAT), a city wide organisation facilitating sustainable activities. c/o LA21 Unit, Room 145, City Hall, Bradford, BD1 1HY. Tel: (01274) 754123

○ Further activities

· Use an urban or highly built up site as one of the options for people to do their design exercise on.

· Look at interactions between the city and countryside and use of the urban-rural edge.

· Look at how elders, adults, adolescents and kids use the city. How could relationships between generations be enhanced?

· Introduce the Pattern Language. Use one or two patterns as examples and show how they can be incorporated into the design process.

○ Link to the ethics and principles

I remind students that all the principles, design methodologies and ethics apply to an urban setting. Nothing is different.

○ Agenda 21 link

Chapters 3, 4, 5, 6, 7 & 10.

○ Further research

A Pattern Language

Urban Permaculture

Sustainable cities, Walter, Arkin (Ed), Crenshaw

Urban Agriculture – Food, Jobs & Sustainable Cities, United Nations Development Programme (UNDP), 1996

Greening Cities, Roelofs, J, The Bootstrap Press, 1996. Many good case studies

Growing Food In Cities Report, Garnett, T, NFA/SAFE alliance, 1997. Full of urban food growing case studies

Cities and Natural Process, Hough, M, Routledge, 1995

Eco-city Dimensions: Healthy Communities, Healthy Planet, Roseland, M, New Society Publishers, 1997

New Economics Foundation, Cinnamon House, 6-8 Cole Street, London, SE1 4YH. Tel: 020 7407 7447 Email: neweconomics@gn.apc.org

Planning for Real, Neighbourhood Initiatives Foundation, The Poplars, Lightmoor, Telford TF4 3QN. Tel: (01952) 590 777

Contact your local Agenda 21 officer for local projects, contacts and initiatives.

Urban Oasis, Propeller Productions, 1996 (video)

Trading Skill

Andy Langford

Why I include this session

Many of us experience employment as an unsatisfactory exchange in which the rewards often do not compensate the accompanying stress. Also employment can mean working at tasks which maintain the current destructive economic systems and help to despoil the earth. To increase opportunities to make an ethical living we may choose to opt for self-employment rather than employment.

Will Hutton's 40:30:30 analysis, given in The State We're In (Ref 1), of the livelihood patterns in our society is useful here. This analysis places people in lifestyle/earning categories of employed (40%), self-employed and short-term contracts (30%) and on benefit or low paid (30%). People are often surprised to see that the majority of people are not employed.

Objective

To explore the skills required for self-employment.

Learning outcomes

By the end of the session, students will be able to:
· List essential TradingSkills.
· List which skills they already have, and which they need to acquire before venturing into self-employment.

Context

TradingSkill comes with other material on sustainable livelihoods which includes:
· Growing, making and selling produce, ditto goods and services so as to create localised markets; running a small business in effect.
· Designing and operating trading infrastructure that enables the above, including trading, banking and credit systems.
· Any description of the world wide permaculture network, in which self-employment and voluntary effort predominate over employment.

Duration

75 minutes

○　How I teach this session

This topic is usually programmed although it may be opened in a Matters Arising session. (See 'Patterns for Course Design – An Open Agenda Approach', pages 32-34.) So it may either be prepared for or off the cuff. Either way it will be different each time according to the course group. Here I describe one way it might go as a prepared session.

Introduction – talk

10 minutes

Why I teach this topic – as above.

Brainstorm

10 minutes

Ask people to name some ways people make a living by trading on their own behalf and identify trades in which self-employed people predominate.

Interactive talk

35 minutes

1 Describe the theory

People who thrive in self-employment have a specific set of skills and attributes which they have learnt and acquired. These are largely invisible to people used to employment or unemployment.

　　Without TradingSkill running a business is much more risky!

2 TradingSkill attributes

· Persistence:
   · the ability to work at realising a vision when there are no quick gains to be had.
   · willingness to work at the less interesting but essential tasks of trading.
· Empiric-ness:
   · willingness to let go of strategies that aren't working.
   · may mean modifying beliefs to accommodate the practical results of activity.
   · involves complex judgements about when to apply persistence and when not to.
· Casino-istic:
   · the ability to distinguish between a good risk or investment and a bad one.
   · the ability to find alternative uses for investments which go wrong, eg premises or equipment.
· Sensate learningness:
   · a preference for learning by doing, hands-on.
   · attention to detail.
· Number skills:
   · understanding how money and stock flow through a business.
   · how to keep track of these flows and manage them using a variety of ratios.
   · understanding business language, eg cash flow, balance sheets.
   · having enough maths.

3 Means of acquiring TradingSkill

Listed in my order of preference:

· brought up in a family that makes a living through TradingSkills.
· Start a low investment business in your teens and gain experience.
· Work as an apprentice with someone who is skilled and agrees to teach you.
· Start a business – making mistakes is a great way to learn but make the mistakes small enough to avoid ruin.
· Go to business classes and learn the theory – but note that practical experience is still needed.

Activity 1

10 minutes

Invite students to assess their TradingSkills in pairs, one at a time.

Activity 2

10 minutes

Ask people who live by TradingSkills to list any other attributes they consider important and tell how they acquired their own skills.

Graphics

Draw a mind-map of TradingSkill during the session.

○ Living examples

Myself. My family were all teachers and thus I had no trading at all in my background. I acquired my TradingSkills by surviving an early business disaster, brought on by a complete lack of consciousness that there were any skills required. It was 10 years after this hard fall before I was ready to give trading another go, and then only after two years of business education, much of which was irrelevant as it was designed for managers of large corporations. Now I have lived by my TradingSkills for many years.

I can illustrate each of the attributes with examples drawn from several business experiences. I count TradingSkills as equally valuable as permaculture design skills and good people skills.

○ Agenda 21

Chapter 3.

○ Bibliography

1 The State We're In, Hutton, W, Vintage, 1996

○ Further research

Honest Business, Phillips, M, & Raspberry, S, Random House, 1981
Marketing Without Advertising, Phillips, M, & Raspberry, S, Nolo Press, 1998
The Path of Least Resistance, Fritz, R, Butterworth Heinemann, 1994

Me, My Greatest Asset!

Andy Goldring

Why I include this session

I believe that one of the greatest limiting factors in society at present is a lack of personal and community self-confidence. This session helps students to realise that they have a wide range of talents and skills which they can utilise for personal and community benefit.

Objective

To give students the opportunity to evaluate their own skills and abilities, and thus raise self-confidence.

Learning outcomes

By the end of the session students will be able to:
· Identify current skills.
· Identify areas of future personal development.
· Help others to build up self-confidence and self-esteem.

Context

This activity fits well within a larger session focusing on individuals or Zone oo. This activity also fits well with a session on LETS.

Duration

In the classroom: 60 minutes
Homework: up to 90 minutes

How I teach this session

Introduction
5 minutes
I show the different charts that will be used and what we will achieve by the end of the session. (See example – charts below.)

Filling in the charts (Ref 1)
50 minutes
· My Abilities
· My Transferable Skills with People
· My Transferable Skills with Things
· My Transferable Skills with Ideas, Information & Data

To fill in the charts fully takes a couple of hours and deserves the time required. So during this session it is not important to complete them all, but rather to make sure everybody knows how to fill them in. However, focusing on My Abilities and one of the Transferable skills with... charts, allows students to get some tangible results before the end of the session. I set the completion of the charts as a piece of homework.

This is a good opportunity for the development of interpersonal and intrapersonal skills, so it is well to remind students of skills they learnt during sessions on listening.

Students can do this session by themselves, with the teacher moving between them to stimulate their memories and prompt new thinking if required. Students can also pair up if they wish and interview each other.

Closing thoughts

5 minutes

Towards the end of the session, students can be invited to share something wonderful they have discovered or observed about themselves.

○ Further activities

See the PMI session (pages 117-121), which can be used by students to evaluate themselves, the course, tutors, venues, designs or any other matters! During a course there is plenty of time for each student to become confident with using this thinking tool, and in combination with the filling in of the charts developed by the Human Scale Development Initiative, the PMI can provide a powerful way for students to evaluate themselves confidently and in a balanced way (Ref 1).

○ Bibliography

1 Contact 'HSDI: A Human Scale Development Initiative', 8b Vicars Road, Leeds, LS8 5AS. Tel: (0113) 240 0349/(0113) 262 2268, for the charts mentioned in this session

○ Further research

What Colour Is Your Parachute?, Bolles, R N, Ten Speed Press, updated annually. Contains useful charts which could be used if the HSDI charts are not available

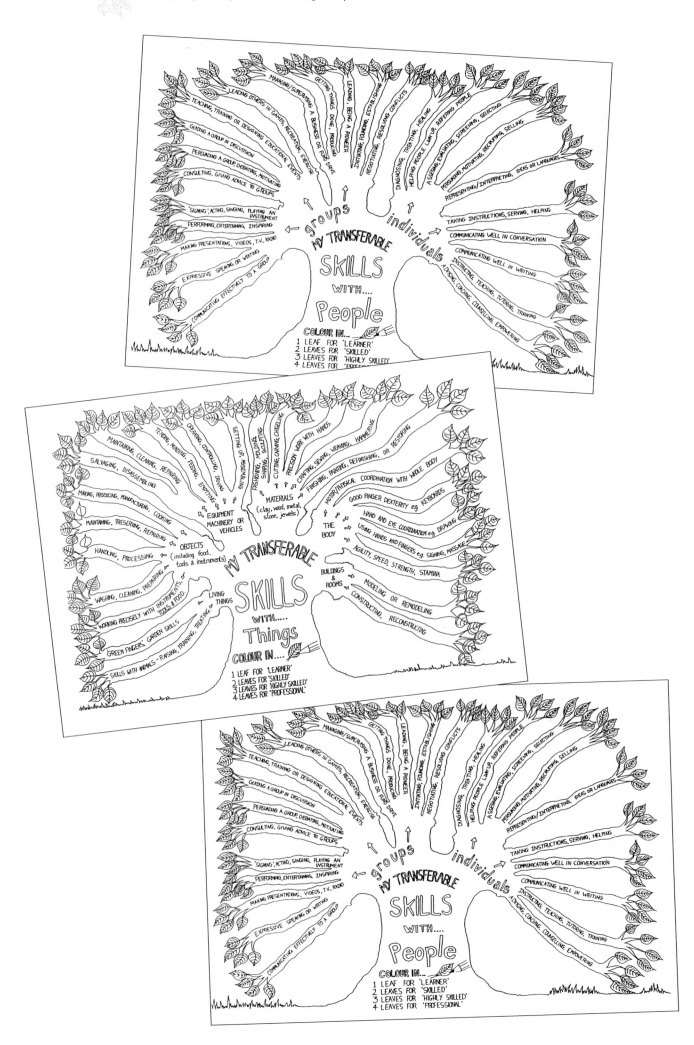

⇨ Gaia Healing Visualisation

Cathy Whitefield

○ **Why I include this session**

Because I want to bring a spiritual energy into the course and into people's lives in an accessible way.

○ **Objective**

To enable students to experience the spiritual energy which can be generated in a group, and then be used for healing purposes. In this case it is used to help heal the planet.

○ **Learning outcomes**

By the end of the session students should have an idea of the energy which can be created in a group and how it can be focused to a positive purpose.

○ **Context**

At any time which feels appropriate. For example I have used it at the opening of the second week on a two-week course, and as part of the ending on both two-week courses and courses run as a series of weekends.

○ **Duration**

5 -10 minutes

○ **How I teach this session**

The people sit or stand in a circle, holding hands.

Imagine a six-pointed star of light above your head. Visualise its rays coming down through your crown and resting as an orb of silvery healing energy in the heart centre (mid point of the chest between the breasts). Focus your attention and breath on this area. Imagine this energy growing stronger and bigger with each in-breath, and with each out-breath imagine it beginning to move down your arms and energise your hands.

Now be aware of the hands you are holding on either side of you, and allow the healing energy to come out through your right hand, and be passed round the circle as you in turn receive energy back through your left hand. Recharge this flow with the healing energy coming down from your heart.

Allow the energy to circulate for a minute or longer as appropriate.

Now visualise the Earth as seen from space in the middle of our circle, this beautiful, blue, revolving planet. Gently let go of your partners' hands, rub your hands together, and hold them in a comfortable position, palms facing forward. Then consciously send the healing energy we have generated together towards the visualised Mother Gaia, surrounding her with light and love.

Hold this phase for however long you feel is right for the group, perhaps two or three minutes, in silence.

Now we will begin to come out of the visualisation. Begin to take some deeper breaths, gently becoming aware of yourself back in the room, opening your eyes and beginning to move and stretch as you feel ready.

○ **Further research**

Creative Visualization, Gawain, S, Eden Grove Editions, 1988

Designing for Human Health

Krysia Soutar

Why I include this session

· To draw awareness to the crisis of the deterioration in human health on a local and global scale, and show the connections with the current ecological crisis.
· To highlight the knock-on effects of unnatural food production and the growth of unsustainable health care systems.
· To show that by applying our understanding of permaculture design we can begin to create the foundations for a future which works with nature not against it, for the well-being of all living things, including ourselves.

Objective

To show how permaculture design can be applied to health care systems to improve quality of life and sustainability.

Learning outcomes

By the end of the session, students will be able to:
· Identify at least five features of the current approach to health care provision which are unsustainable.
· Discuss the benefits of a permaculture design approach towards improving quality of life.
· Relate human needs to the availability of natural resources.
· Select design elements which can be used to meet human needs, sustainably.

Context

After principles and design methodologies.

Duration

The four themes below can be covered in one session of 80 minutes. Another session will allow participants to explore the subject in more detail and cover the further activities.

How I teach this session

1 Current health issues in the developed world – *whole group brainstorm, or small group discussion and feedback*
20 minutes
· After an initial brainstorm, or discussion in small groups, outline diagrammatically the current situation of people care in the modern world, emphasising health issues.
· Remind participants that 'the problem is the solution'.

2 Visioning for a permaculture health care system – *assign participants to small groups, then feed back*
20 minutes
· Compare and contrast the current health care system with a future permaculture health care system.

Ask students to develop proposals.

· Emphasise that new proposals are to be in addition to, rather than a replacement of, the existing systems. Typical results are shown below.

Types of health care system	
Current system	**Permaculture system**
• national	• bioregional
• centralised	• neighbourhood-based
• big hospitals	• local centres
• technology – surgery, drugs, genetic engineering transplants	• human skills – medicinal plants, foods, physical treatments, diversity
• 'western' (reductionist) science	• interconnectedness • natural cycles
• resources consumed – unsustainable	• resources created within the system
• output pollution	• earth care
• causes ill health (eg drug resistant bacteria, side efects of drug therapy)	• sustains health

3 Designing for human health – brainstorm, facilitate and record

20 minutes

· Use the framework in the table above. You could give out post-it notes or cards and ask students to stick them onto a large chart with the headings at the top.

· Ask students to identify human needs, eg nutrition, living environments, activity etc. Then match with suitable elements and available resources. (See mind-map.)

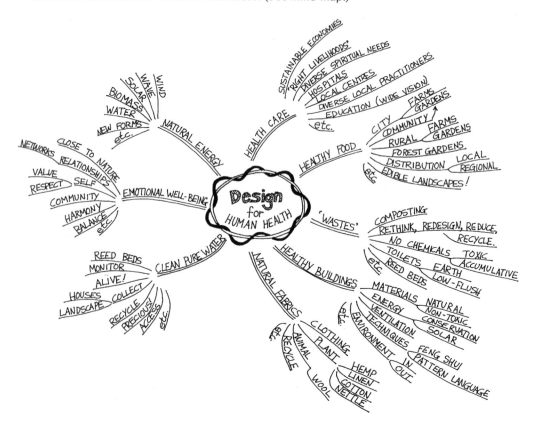

4 *Human health as a multiple output* – *brainstorm, facilitate and record as a mind-map*

20 minutes

· Look at examples of how existing permaculture designs contribute towards health in a positive way.
· Show how 'one element can serve many functions' – identify the health care functions.
· Outline some design approaches. (See below.)
· Discuss implementation and maintenance of new healthcare systems. In outline, focus on next steps.

Human needs (function)	Elements that service the function	Resources and design considerations
Healthy food	Food production • vegetable • animal, mineral	Healthy soil, perennial plants, local food links etc
Elimination of wastes	Reed beds, compost toilets, grey water systems	Plants to break down waste and microbes Design for safe and sustainable processing
Shelter	Buildings, clothes, heating	Local materials, energy efficiency, siting of affordable housing, self-build

○ Further activities

· Practical exercises experiencing body energy, eg breathing, Japanese Do-In.
· Outline techniques and bodies of knowledge for natural health care from other cultures, past and present.
· Look at patterns in nature which link humans to the rest of the natural world. Show examples using charts or slides.

○ Agenda 21 link

Chapter 6.

○ Further research

Nature Doctor, Vogel, Dr H, Mainstream, 1990
The Book of Ferment and Human Nutrition, Mollison, B, Tagari, 1993
Quantum Healing, Chopra, D, Bantam, 1990
The Book of Macrobiotics, Kushi, M, Japan Publications, 1977
Healing Wise, Weed, S, Ashtree Publishing, 1989

Making Connections or "What's My Role in Life?"

Jamie Saunders

Why I include this session

A session reinforcing the connections between course material and their own learning experience helps students' personal development as permaculture designers. In particular, this session looks to support the student in self-belief, and in confidence that they know 'how to do permaculture', and how they might set about improving their design skills.

Objective

To support individuals in developing their own understanding of and contribution to sustainability through permaculture design.

Learning outcomes

By the end of the session students will be able to:

· Appreciate the range of opportunities for promoting and implementing sustainable design.
· Explain where they see their key skills and interests fitting in to the development of long-term quality of life.
· Develop designs for their lifestyles including working life, which take account of their 'internal drives' and their impact on other people and communities.
· Appreciate what makes an effective permaculture designer and build this knowledge into their design work.

Context

About three-quarters of the way through the course.

Duration

60 minutes

How I teach this session

The essence of the session is to draw out of the individuals in the group:

· Their understanding of permaculture.
· How they are applying or intend to apply it to their own lifestyles.
· Their role in communities and organisations.

Brief introduction
5 minutes
Aims of the session, as outlined above.

Brainstorm (Consider all factors, what is missing?)
30 minutes

What makes an effective permaculture designer?

I map points on a flipchart, and group them into themes and issues. In most instances, these cover issues such as:

· Commitment to the ethics and principles of permaculture/sustainability.
· The need for ongoing personal development and lifelong learning. (The course as a taster and initial framework for a lifetime's effort.)
· The model of Succession as an inspiration for permaculture design practice.
· Process and project management skills and applying them to life choices, ie planning our future as opposed to planning a holiday.
· Tackling local needs at different scales, eg minimum intervention, networking at a regional or national level.
· The role of the permaculture activist as community networker, facilitator and go-between.
· The permaculture designer as generalist or holistic connector – the impact of this on specialist skills and activities.
· The aim of local self reliance (cross-refer back to concepts of guilds, niches, elders and apprenticeship).
· The relationship of permaculture designers to other organisations, groups, professions.
· The progress towards payment for design work, the need to appreciate accountability to clients and community, and liability.
· Confidence, positivity and levity (how did the apple get up there in the first place?) as vital in inspiring and supporting others.
· Incremental design, rolling and back-pocket permaculture. (Little steps in the right direction make a difference but don't forget the bigger picture...)

Break

Personal vision exercise

20 minutes

A3 or A4 (see opposite), and a contribution to the personal portfolio. The contents of this exercise are often personal and the tutor should ensure that individual's rights to privacy are maintained. Encourage sharing, but not at the expense of isolating a student from the group. If a number of views are shared then it should be possible to establish learning sets from within the membership of the course. This can also help in future sessions, for group work or identification of project leaders, apprentice teachers and so on.

This fits very nicely into identifying elders, mentors or opportunities for outreach with the permaculture community or in the wider world. This would also be an early prompt towards portfolio development, Diploma material, and potential livelihood opportunities.

Personal visioning sheet

Name   Date

Where am I now?   Vision

Next achievable steps

Handouts

· A4 Personal Visioning Sheet (see previous page).
· 6 Key Activities for successful implementation of change (Ref 1).

avoid over-organising   *communicate like never before*

ensure early involvement

managing change

develop support networks

gain commitment

turn the perception of threats into opportunities

Source: *Managing change and making it happen*, Roger Plant

· Consultancy Cycle – focus for discussion on role of agents of change

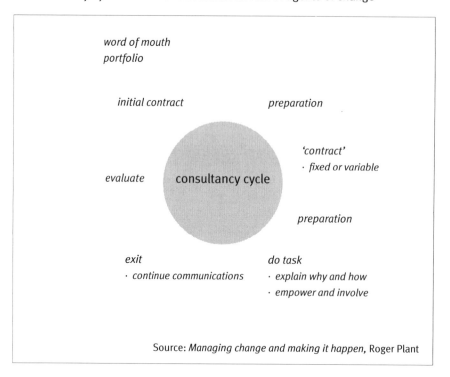

word of mouth portfolio

initial contract   *preparation*

'contract'
· fixed or variable

evaluate   **consultancy cycle**

preparation

exit   *do task*
· continue communications   *· explain why and how*
· empower and involve

Source: *Managing change and making it happen*, Roger Plant

○ Further activities

· Mapping sustainable lifestyles – next steps?
· Livelihood/permaculture design application to different circumstances.
· Portfolio/learning set guide and linkages to Diploma material.
· Targeted learning – depth and breadth.
· Designers Register. See 'About the Permaculture Association', pages 362-365.
· Wednesday's Guardian.

○ **Bibliography**

1 Managing Change and Making it Stick, Plant, R, Gower Publications, 1987

○ **Future research**

Systems, Management and Change: a graphic guide, Carter, R, et al, Paul Chapman Publishing Ltd & Open University, 1984

How to be an even better manager, Armstrong, M, Kogan Page, 1994

Continuous Improvement Tools, Volumes I & II, Chang, M, Kogan Page, 1993

The 80/20 Principle, Koch, R, Nicholas Brearley Publishing, 1998

Facilitators Guide to Participatory Decision Making, Kaner, S, New Society, 1996

Power in our hands, Neighbourhood based World Shaking, Gibson, T, Jon Carpenter, 1996

Putting Power in its place – create community control, Plant, C & Plant, J, New Society, 1992

Age of Unreason, Handy, C, Century Business, 1993

Empty Raincoat: Making Sense of the Future, Handy, C, Arrow, 1994

The Learning Organisation: A Strategy for Sustainable Development, Pedlar, M, et al, McGraw Hill Book Co, 1991

Patterns in Human Behaviour

Andrew Goldring

○ Objectives

To identify patterns in human behaviour, so that a pattern or checklist can be created for use in design work.

○ Learning outcomes

By the end of this session students will be able to:

· List human behaviours.
· Explain why this is useful.
· Go on to develop sustainable strategies for each behaviour.

○ Context

I set this as a piece of 'home research', at the end of the first weekend (on six weekend courses.) It is reviewed and developed on the first morning of the next weekend.

○ Duration

30-60 minutes home research
40 minutes in the classroom

○ How I teach this session

Setting the home research
5 minutes

I say to the group, "List ALL the things that you do. This list will only be seen by yourselves, so there is no need to be shy. When this list has been completed, work through each activity and tick any which you think you have in common with the rest of the human race. Include any male or female specific activities. Try to see if there are any 'higher level' concepts in what is left, eg playing dominoes may be part of 'recreation'. Put all the ticked activities onto a new list and bring it in next weekend."

Review – collating lists and finding overlaps
15 minutes

Invite students to put forward activities from their lists, until all the points have been collated on the board.

I then ask the group to direct me in grouping them. I do this because the same activity may have been given a few times but with different wording. Ask the group if there are any glaring omissions – birthing often gets missed – ask if anyone has ever had kids!

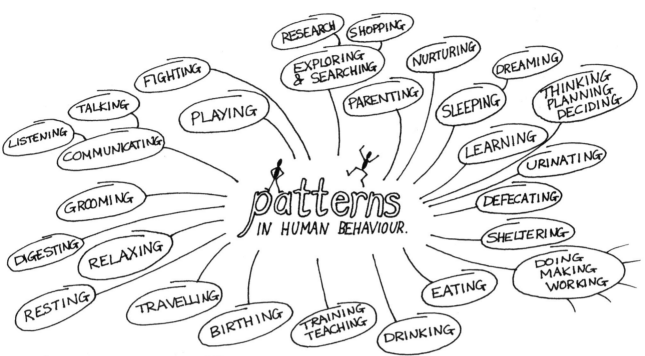

△ TYPICAL RESULTS AFTER COLLATION
○ CONSULT ETHOLOGY BOOKS FOR MORE INFORMATION / BACKGROUND
☆ MANY THANKS TO WOLF WHITE FOR HIS RESEARCH & INSPIRATION.

Discussion in small groups

10 minutes

"In small groups discuss the following question, keeping notes of all points raised. What implications does this list have for sustainable design?"

Whole group discussion

15 minutes

Invite each group to put forward points raised in their discussions. Explore and expand on points as necessary. I close the session by thanking everyone for their excellent work and saying that we will return to this list at a later stage, when looking at bioregions.

Bioregions – Defining Characteristics

Andrew Goldring

Why I include this session

An understanding of how bioregions are defined can help students to recognise their own. By looking at one's own home place in a bioregional context, an appreciation of what is special and precious about it can develop. It also helps to give a context within which we can steadily work to improve things. Working to improve one's own bioregion is less daunting than trying to 'save the planet'. Tasks that seem too big, rarely get done.

Objective

To identify characteristics that define bioregions.

Learning outcomes

By the end of the session participants will be able to:
· Identify their own bioregion.
· Explain the concept easily to others.
· List defining characteristics.

Context

Middle of the course.

Duration

45 minutes

How I teach this session

Preparation
Get a large map, making sure that the bioregion is fully contained within the map, with plenty of overlap with other bioregions.

Opening
5 minutes
I ask someone to act as a scribe for the session.

I then ask the group what they understand by the term bioregion. Splitting the word up helps, as does asking them to consider how communities may have operated many centuries ago.

Exercise
20 minutes
As soon as we have established an adequate understanding of what a bioregion is, I get a pin and place it on the map and ask "Is this in your bioregion?". Whether the answer is yes or no, I ask why. The reason given is noted by the scribe. This continues with quite a few pins, until there are enough to join them up and everyone can get a reasonable idea of the boundaries of the bioregion. Students are encouraged to come and place pins on the map themselves.

To finish the map stage, I ask which of the boundaries are fuzzy, which are hard, and why.

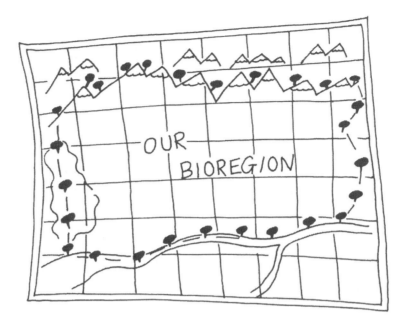

Discussion

20 minutes

The notes made by the scribe are now reviewed. It is pointed out that the group has intuitively come up with the defining characteristics of a bioregion. I spend a few minutes pulling these points together and fill in any gaps there might be. This is followed by about five minutes of questions and answers to cover any last points.

○ Further activities

Using the list generated in the 'Patterns in Human Behaviour' session (pages 316-317), ask the group to name bioregional level institutions, systems and facilities which would be required to develop and maintain a sustainable bioregion.

Link this to the bioregional resource index in the Designers' Manual (Ref 1).

If the course is mainly attended by local people, a further home research exercise can be set to identify local resources and activities which do/could contribute to a sustainable bioregion, thus identifying gaps and opportunities. This information feeds in well to a LETSystem session, and provides students with a local map which can be used after the course.

○ Bibliography

1 Permaculture: A Designers' Manual, pp510-513

○ Further research

A Pattern Language (Pattern for Independent Regions)
Our Ecological Footprint
Bioregional Development Group, Sutton Ecology Centre, Honeywood Walk, Carshalton, Surrey,
SM5 3NX. Tel: (020) 8773 2322
Boundaries of Home, Mapping for Local Empowerment, Aberley, D (ed) (The New Catalyst Bioregoinal Series), New Society Publishers, 1993

Bioregions and Sustainable Settlements

Mark Warner

Objective

To give an overview of bioregionalism and sustainable settlement design, and demonstrate how permaculture design is useful in these contexts.

Learning outcomes

By the end of the sessions students will be able to:
· Begin the processes of mapping their bioregion.
· Begin developing a strategy for co-ordinated land use and resource management, in co-operation with the rest of the bioregion's inhabitants.
· Have an outline understanding of the process of sustainable settlement design, within urban or rural, new-build or retrofit contexts, and in relation to the settlement's wider bioregional development.
· Know where to go for further information and training in sustainable settlement and eco-village design.

Context

By nature this subject includes all other subjects taught on a permaculture design course. It should be introduced at an early stage to give the framework in which to fit the component subjects. (See Phases 1 and 2 below.) A more in-depth study of the subject follows near the end of the course. (See Phases 3 and 4.)

Duration

Phase 1: 50 minutes
Phase 2: 60 minutes
Phase 3: 150 minutes
Phase 4: 20 minutes

Total duration is 4 hours 40 minutes.

How I teach this session

Phase 1

Short lecture

20 minutes

Along the lines of: Sustainable settlements are human scale communities combining the individual, society and economy, and the environment, within the context of the surrounding landscape's natural boundaries of geology, hydrology and ecology. Therefore it is necessary to co-ordinate the management of the region's resources, such as land, water, energy and people, to ensure the greatest benefit for all beings within the catchment area. When considering a proposal for the development of a site it must be seen within the bioregional context or it is unlikely to be successful, or at least be unable to reach its full potential in the long term.

Regional sustainable development requires a strategic approach, taking into account the needs of the landscape and the people living in it. We are delving into the world of planning and local authorities. The constant need for new housing and economic development that is, at present, being provided for on an unsustainable basis has to be provided for through good design.

Discussion

30 minutes

Bring together aspects of settlement design, community development and regional resource management. Handouts giving an overview of Bioregionalism and Sustainable Settlements are circulated.

Phase 2

Visioning exercise

45 minutes

What kind of community would people like to live in?

This can be done through drawings, small group discussions, short 'plays' or 'TV show from the future' type activities. Give the groups 30 minutes to prepare and 15 minutes to present. Point out the diversity of visions people come up with, but also point out the overlaps and common ground.

Outline of component subjects

15 minutes

Establish the framework for the component subjects which will be covered in the middle part of the course.

Phase 3

Interactive review

20 minutes

Having covered the component subjects they can be reviewed within the context of the Bioregions and Sustainable Settlement framework.

Case studies

40 minutes

Take a few in-depth looks at some prime examples of, and working models for, Bioregions and Sustainable Settlement developments. This can be done with the aid of case study handouts, slides, lecture and discussion.

Exercises

1 Bioregional mapping

45 minutes

With the whole group, using hydrological, ecological, cultural, and other criteria to identify the local bioregion, define the area on a small scale map (make sure it covers a large enough area), taking note of things like urban/rural areas, infrastructure, landscape, etc. Repeat this process on a larger scale map for the local sub-region and begin to record more detailed information from a selection of local data sources and from the participants (see 3 below). Students should be encouraged to think creatively about information recording methods and presentation.

2 Ecovillage design

45 minutes

In groups of three, ask the students to outline the processes for designing a small sustainable settlement on some land in the local sub-region in terms of environment, economy and society. Students will be able to make use of the bioregional maps produced in the previous exercise and should demonstrate some ways that their settlement design relates to the bioregion and sub-region.

3 Planning for real (if not covered elsewhere)

Phase 4

Information

20 minutes

· A general handout listing useful publications and organisations.
· Contacts for bioregional and/or sustainable settlement development in the local region.
· Information on any upcoming eco-village, sustainable settlement or bioregional design courses.
· A step-by-step guide to making successful planning applications, and how to influence strategic development and management within your bioregion and community.

O Further activities

· Talks on co-housing, co-operatives or sources of funding.
· A more in-depth discussion about legal structures.
· Invite a friendly planning officer to talk about how the planning system works.

O Living examples

· Acorn Televillages Ltd, The Televillage, Crickhowell, Powys, NP8 1BP. Tel: (0800) 378848. Email: 100273.2505@compuserve.com
· Hockerton Housing Project, The Gables Workshop, Hockerton, Southwell, Nottingham, G25 0PP. Tel: (01636) 815614. Email: nwhite@fatmac.demon.co.uk
· Tinkers Bubble, Little Norton, Stoke-sub-Hamdon, Somerset, TA14 6TE.
· The King's Hill Collective, Cockmill Lane, East Pennard, Shepton Mallet, Somerset, BA4 6TR.
· Centre for Alternative Technology, Machynlleth, Powys, SY20 9AZ. Tel: (01654) 702400.
· Findhorn (Ecovillage Project), The Park, Findhorn Bay, Forres, Morayshire, Scotland, IV36 0TZ. Tel: (01309) 690956.

For up-to-date discussions, information and contacts for living examples check out:
· The Bioregional discussion list on the internet at:
  http://csf.colorado.edu/bioregional
· The ecovillage websites at:
  http://www.gaia.org/uk
  http://www.gaia.org
· For more urban based studies: http://www.urbed.co.uk/sun

○ **Further research**

The Permaculture Plot

Diggers and Dreamers 1998/9, Diggers and Dreamers Publications, 1997#

Low Impact Development, Fairlie, S, Jon Carpenter, 1996

Village Homes' Solar House Designs, Bainbridge, Corbett & Hoffacre, Rodale, 1979

Dwellers in the Land – A Bioregional Vision, Sale, K, New Society Publishers, 1991

Permaculture and Planning, Rob Hopkins, Self-published Thesis

Low Impact News, L.I.N., c/o 71 Dale St, York, YO2 1AE. Tel: (01904) 647 235

Action for Sustainable Rural Communities, c/o Lowe Rae Architects, Three Crowns Yard, Penrith, Cumbria, CA11 7PH. [contact Rod Hughes] Tel: (01768) 890067. Email: lowe.rae@dial.pipex.com

Eco-Village Network Email: evnuk@gaia.org

Global Ecovillage Network, Skyumvej 101, 7752 Snedsted, Denmark. Tel: (00 45) 97 93 66 55. Email: gen@gaia.org.

Ecological Design Association, The British School, Slad Road, Stroud, GL5 1QW. Tel: (01453) 765575.

Association for Environment Conscious Building (AECB), Nant-y-Garreg, Saron, Llandysul, Carmarthenshire, SA44 5EJ. Tel: (01559) 370908.

Radical Routes, Cornerstone Housing Co-op, 16 Sholebroke Avenue, Chapeltown, Leeds, LS7 3HB. Tel: (0113) 262 9365. Email: cornerstone@gn.apc.org

HSDI: A Human Scale Development Initiative, 8b, Vicars Road, Leeds, LS8 5AS. Tel: (0113) 240 0349.

The Natural Step UK, Thornbury House, 18 High Street, Cheltenham, Glos, GL50 1DZ. Tel: (01242) 262744.

Planet Drum Foundation, PO Box 31251, San Francisco, California 94131, Shasta Bioregion, USA. Tel: (00 1) 415 285 6556. Email: planetdrum@igc.apc.org

Living Village Trust, The GEN Centre, Church Street, Bishops Castle, Shropshire, SY9 5AA. Tel: (01588) 630122. Email: living.village@btinternet.com

Community Patterning

Chris Dixon

Why I include this session

Sustainable agriculture and sustainable communities are mutually inseparable; we cannot have one without the other.

In designing sustainable systems we undertake careful observation of a site, noting such things as existing resources and what is working well already. In the same way we can observe and study existing and past communities, searching for clues, resources or patterns which we may use in the modification of existing communities or the design of new ones.

Community Patterning provides a very objective and tangible way of approaching or working with the People Care ethic.

Objective

To demonstrate the use of patterning as a tool for understanding and designing communities.

Learning outcomes

By the end of this session students will be able to:
· Explain that community exists in the minds and activities of its members.
· Recognise similarities between patterning in environments and communities.
· Identify traditional connections within a community.
· Generate further events and activities which can draw members of a community together.

Context

This topic can be usefully taught in the second half of a design course. Some prior experience and understanding of patterning in Earth Care systems is useful to demonstrate the similarities of approach, the fuzzy nature of the two sets, Earth Care and People Care, and the use of patterning as a linking discipline.

Duration

90 minutes

How I teach this session

Introduction – talk

5 minutes

I begin by saying how the bulk of this work has been developed from the observation of and interaction with the traditional local communities of the Mawddach estuary, in particular Abergeirw. I am indebted here to many members of that community, Nesta Wyn Jones in particular. Remnants of traditional patterning can be found throughout Britain. We may also include tribal systems and prehistoric evidence.

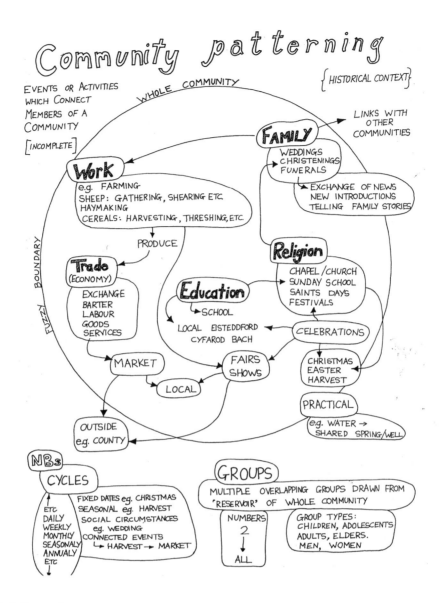

Community patterning

EVENTS OR ACTIVITIES WHICH CONNECT MEMBERS OF A COMMUNITY

[INCOMPLETE]

{ HISTORICAL CONTEXT }

WHOLE COMMUNITY

LINKS WITH OTHER COMMUNITIES

Family
WEDDINGS
CHRISTENINGS
FUNERALS

EXCHANGE OF NEWS
NEW INTRODUCTIONS
TELLING FAMILY STORIES

Work
e.g. FARMING
SHEEP: GATHERING, SHEARING ETC.
HAYMAKING
CEREALS: HARVESTING, THRESHING, ETC.

PRODUCE

Religion
CHAPEL/CHURCH
SUNDAY SCHOOL
SAINTS DAYS
FESTIVALS

Trade
(ECONOMY)
EXCHANGE
BARTER
LABOUR
GOODS
SERVICES

Education
SCHOOL
LOCAL EISTEDDFORD
CYFAROD BACH

CELEBRATIONS

MARKET

FAIRS
SHOWS

CHRISTMAS
EASTER
HARVEST

LOCAL

PRACTICAL
e.g. WATER →
SHARED SPRING/WELL

OUTSIDE
e.g. COUNTY

BOUNDARY   FUZZY

NBs
CYCLES
ETC
DAILY
WEEKLY
MONTHLY
SEASONALY
ANNUALY
ETC

FIXED DATES e.g. CHRISTMAS
SEASONAL e.g. HARVEST
SOCIAL CIRCUMSTANCES
e.g. WEDDING
CONNECTED EVENTS
↳ HARVEST → MARKET

GROUPS
MULTIPLE OVERLAPPING GROUPS DRAWN FROM 'RESERVOIR' OF WHOLE COMMUNITY

NUMBERS
2
↓
ALL

GROUP TYPES:
CHILDREN, ADOLESCENTS
ADULTS, ELDERS.
MEN, WOMEN

Guided discussion

45 minutes

I have a pattern relating to the session (see mind-map).

I lead the group through an examination of the connections within a fairly traditional British community of say 100 years ago.

I use a combination of presentation and collation from the group to generate the pattern on the board. Additional points that arise during the session can be included if the group consider them relevant. I begin by drawing a large circle to represent the community in a broad sense, and then add the keywords to the circle as they are generated by the group.

I find it helpful to be clear that I am asking for examples of events or activities which draw members of that community together and that we are thinking of a historical community.

The essential point made here is that a community exists in the minds and activities of its members. Each activity which draws members of the community together will reinforce the experience of community. The experience of community may be felt as creative and beneficial or as destructive and negative. As permaculture designers we would seek to maximise activities which reinforce community in a positive way.

Thinking exercise in pairs (1)

10 minutes

Which connecting activities have been lost or degraded? We swap time in pairs, with each having two minutes to talk, whilst the other actively listens (see 'A Practical Exercise in Swapping Time', pages 142-143). We then mark the changes on the pattern.

Thinking exercise in pairs (2)

15 minutes

How can we design new connecting activities into existing or new communities? We swap time (as above) on this first, and generate a new pattern which will include elements from the first pattern.

In summary

10 minutes

We identify the traditional connections within a community, notice which modern culture has been removed or damaged and move on to consider how to re-energise degraded connections or create new activities which can provide alternative connecting points, such as veg box schemes, LETSystems etc.

Review and feedback

5-10 minutes

Ask students for any observations about the session process, and closing remarks.

At the end of the session I make a point of being obvious about adding new information to my original pattern. The mind-map/session plan is thus an ongoing process subject to review.

○ Link to ethics and principles

Work With Nature, Not Against It. As with land-based systems we may apply careful observation followed by steering in an increasingly productive direction. Humans are, by their very nature, co-operative and tend to live in communities.

○ Further research

Grooming, Gossip and the Evolution of Language, Dunbar, R, Faber and Faber, 1996. A fascinating account of the evolution of language and social groups, providing some excellent background material

⇨ Communities

Cathy Whitefield

○ Why I include this session

Many of us want to live in a better functioning community than the one we presently live in. If we understand what makes communities work we can create what we need more easily. This session helps us both to clarify our needs and wants in our living situation, and to realise them.

○ Objective

To empower students to improve the community where they live, or to find or create a better one.

○ Learning outcomes

By the end of the session students will be able to:
· Explain what makes communities work.
· Reflect on the kind of community which would suit them best.
· Identify their next achievable step in bringing it about.
· Use the listening skills learned earlier in the course, and during the session, with more confidence.

○ Context

This session should be early enough in the course that it can help to build community within the course itself, but far enough into it that people feel reasonably comfortable with each other, and so can make the most of it. On a two-week course it is often the last session of the first week.

○ Duration

45-75 minutes

○ How I teach this session

Opening
5 minutes
First I define community. I do not necessarily mean intentional communities or communal living, though these are included. A community could just as well be a village, urban street or neighbourhood.

Talk
10 minutes
The Scott Peck model of community development (Ref 1).

Brainstorm
25 minutes
What makes communities work? Discuss the brainstorm and fill in any gaps.

The session can end here, but usually I continue with:

Paired discussion

15 minutes

Students choose partners, and each speaks for seven minutes while the other actively listens, using the skills learned in the 'Listening to the Landscape' session. The subject for the first five minutes is My Ideal Community. This could be where they live now, made to work better as a community, or somewhere else. During the last two minutes the subject is My Next Achievable Step.

I tell the group when five minutes and seven minutes is up, and remind them of the change of subject. If I see people discussing, rather than one speaking and the other listening, I remind them what they should be doing.

Ending

5 minutes

If it feels appropriate and there is time, I ask if anyone has anything arising from the pair work which they would like to share with the whole group.

Otherwise I may end in a circle with a minute's silence, holding hands if appropriate. Or I may simply ask the group to look around and acknowledge each other with eye contact before ending the session.

○ Bibliography

1 The Different Drum, community-making and peace, Peck, MS, Rider, 1988

○ Further research

Ancient Futures, learning from Ladakh, Norberg-Hodge, H, Rider, 1991. Depicts an example of a community that worked (video)

Community and Sustainable Development, Warburton, D (Ed), WWF-UK/Earthscan, 1998

Community Building in Britain, Peter Cooper, 1 Evergreen Close, Woolmer Green, Knebworth, SG3 6JN Tel: 01635 47377. Organise workshops on the Scott Peck model of community building

Community Economics

Simon Pratt

○ Why I include this session

It is difficult to get anything done without money, so some understanding of how the money system works and the alternatives available is important.

○ Objective

To explain how the money system works and the alternatives available.

○ Learning outcomes

By the end of this session students will be able to:
· Explain why trading locally is important.
· Know where to find sources of ethical finance.

○ Context

Generally towards the end of the course, although I have successfully included it on introductory weekends.

○ Duration

Two 40 minute sessions

○ How I teach this session

Short presentation followed by exercise and discussion
The subject is very emotive for many people, so the challenge is how to limit discussion while conveying the main points and encouraging participation. I have sometimes found it difficult to limit discussion and have failed to move onto other key points.

Some useful material is available on video, either as a short (5-10 minute) slot for variation, or as an evening activity (Refs 1 & 15).

Opening remarks
· Environmental and social accounting are as important as financial.
· Eg the Body Shop, who are now planning to integrate their social, environmental and financial accounts into one annual report.
· Ecological Footprint – refer to Design Methods and Processes session.
· Social auditing – information from New Economics Foundation which hosts the European Social Audit Institute and is setting up a separate social audit wing (Ref 2).
· Tribal systems of exchange do not rely on money.

Three uses of money
a Store of value.
b Measure of value.
c Commodity, can be traded and speculated.

There is many times more money circulating than is needed, but it's not in the right places. Potential for lots of discussion, move swiftly on to:

Whole group exercise

Matrix of ownership against resources (graphic)

	Ownership	
	local	**outside**
Resources outside	eg market gardens	fishing mining
Resources local	local shops	supermarkets petrol stations

Ask for examples of:

· Outside ownership, outside resources (supermarket).
· Local ownership, local resources (market gardens).
· Outside ownership, local resources (fishing, mining).
· Local ownership, outside resources (local shops).

Follow up by comparing money spent in the local economy with money spent in the multinational economy as a whole group exercise. (See 'Local Food Links', pages 266-269 where this is reproduced in full). In the local, or multiplier, economy, about 80% of each transaction stays local and 20% goes outside. The proportions are reversed in the multinational economy. This demonstrates clearly the importance of spending money locally. Something we can all do more of.

Compare money with water. It's not the total amount entering a community which counts, but the number of uses or duties and cycles of use.

I generally continue with a discussion of Margrit Kennedy's ideas and Silvio Gesell's circulation fee and the famous example of the Austrian town of Worgl (Ref 3, p199 & ref 14). Margrit's four misconceptions about money can be helpful or confuse people totally, so make sure you understand them before attempting to teach! They are:

· There's only one type of growth.
· We pay interest only if we borrow.
· We're all equally well off in the present monetary system.
· Inflation is an integral part of free market economies.

LETSystems

The LETS trading game can be used as a 'market place' exercise and/or earlier in the course as a way to facilitate offers and requests within the group. Participants spend a few minutes writing down their marketable goods and services plus their wants and are then encouraged to trade as much as possible with other participants.

Most successful LETSystems are linked with a locally controlled financial institution to improve local use of and access to the external currency. I then review the various alternatives, encouraging group participation, and provide a few inspiring examples.

Credit unions

It's not so easy to set one up in Britain. You need 11 officers to run the organisation and you have to follow bureaucratic procedure. They are owned and controlled by members, but many do not invest ethically. There are over 500 in this country. They need a 'common bond' of employment, residence, other association. Members can borrow at preferential rates, but only if they are already savers. Popular in Ireland, Canada, Australia, US and New Zealand. Contact ABCUL (Refs 4 & 5).

Ethical investment

For the saver, refer to Peter Lang's book (Ref 6) for where to put your money. Up to now most institutions have used negative criteria for ethical investment, eg no alcohol, bombs, or gambling, but we are now starting to see positive criteria being used, eg energy efficiency. Sources of money for projects, from the 'social economy':

· Radical Routes – loan fund for housing and worker co-operatives (Ref 7).
· Triodos Bank (formerly Mercury Provident) (Ref 8).
· Ecology Building Society (Ref 9).
· ICOF (Industrial Common Ownership Finance) – loans for worker co-operatives and community businesses (Ref 10).
· Shared Visions – loans for permaculture and allied projects (details from Andy Langford via Ref 11).

○ Living examples

· Radical Routes
  · Brambles housing co-op in Sheffield.
  · Organic Roundabout organic food distribution in Birmingham (Ref 7).
· Grameen Bank – set up in 1977 to provide small loans in Bangladesh, unique group security system gives very low default rate (Ref 12). Being duplicated in many other countries, including Aston Reinvestment Trust in Birmingham (Ref 7).
· Maleny, small town in Australia – LETSystem with 800 members and a monthly turnover of 25,000 units, credit union with 3,000 members and assets of 9.5 million Australian dollars. Described by Jill Jordan in Ref 13.

○ Agenda 21 link

Chapter 3.

○ Bibliography

1 Global Gardener video, example of community economics in Germany

2 New Economics Foundation, Cinnamon House, 6-8 Cole Street, London, SE1 4YH.
Tel: (020) 7407 7447. Email neweconomics@gn.apc.org

3 Interest and Inflation Free Money, Kennedy, M, Permakultur Institut, 1988

4 Association of British Credit Unions Ltd (ABCUL), Unit 307, Westminster Business Square, 339 Kennington Lane, London SE11 5QY. Tel: (020) 7582 2626.

5 National Federation of Credit Unions, Unit 1.1 & 1.2, Howard House Commercial Centre, Howard Street, North Shields, Tyne & Wear, NE30 1AR. Tel: (0191) 357 2219

6 Ethical Investment: A Saver's Guide, Lang, P, Jon Carpenter Publishing, 1997

7 Radical Routes (housing co-ops and workers co-ops)

Financial Investments: 28 Hampstead Road, Hockley, Birmingham, B19 1DB. Tel: 0121 551 1132
Email: radical@globalnet.co.uk

8 Triodos Bank, Brunel House, 11 The Promenade, Bristol, BS8 3NN. Tel: (0117) 973 9303

9 The Ecology Building Society, 18 Station Road, Cross Hills, Nr. Keighley, West Yorkshire, BD20 7EH.
Tel: (01535) 635933

10 ICOF, 115, Hamstead Road, Handsworth, Birmingham, B20 2BT. Tel: (0121) 523 6886.

11 Sources of funding for permaculture projects are available from the Association office. (Useful as a
handout.) BCM Permaculture Association, London, WC1N 3XX. Tel: (07041) 390170

12 After the Crash, Dauncey, G, Green Print, 1988

13 Eco-Villages and Sustainable Communities, report from 1995 conference, Findhorn Press

14 The Living Economy, Ekins, P (Ed), Routledge, 1986

15 Manfred Max-Neef video, part of Visionaries series

Further research

Short Circuit

Community Works, booklet produced by the New Economics Foundation. A good short summary of
practical community economic strategies

'Investment and Ethics', Issue 15, Permaculture Magazine, Permanent Publications, 1997

See 'LETS Trading', pages 336-337, for list of LETS organisations.

LETS Trading

Angus Soutar

Why I include this session

Teaching about LETS introduces a useful tool for individual and group empowerment and capacity building. It also provides a useful case study in applying design principles to social and economic situations. Although he had not studied permaculture, the principles that Michael Linton used in the initial design work are identical to those employed by permaculture designers. The same applies to the work of Luis Lopezllera on the Tlaloc currency system (Ref 1). Something very much like LETS will be required to support the truly local and bioregional sustainable economies of the future.

Objective

To give an example of the application of permaculture design to community-based financial systems with the aim of promoting sustainable development.

Learning outcomes

By the end of the session students will be able to:
· Recognise the nature of the problem that LETS seeks to address.
· Identify the permaculture principles that apply to the LETS design.
· Illustrate how a LETS trade works.
· State the difference between a commitment and a debt.
· List organisations that can provide further training and support.

Context

This module may be included at any point in the course, although best grouped with related topics.

Duration

As part of a community economics or community building session: 5-10 minutes
As outlined below: 45 minutes
For a group who are serious about implementation: two 45 minute sessions

How I teach this session

Interactive talk, drawing from the group's knowledge
Start with the problems with money (see below), then show how LETS is a designed solution. Illustrate the permaculture principle – the problem is the solution.

Problems with money – moves almost anywhere – has to be kept in scarce supply – centrally issued – no local responsiveness, 'they' do it. (See illustration on next page.)

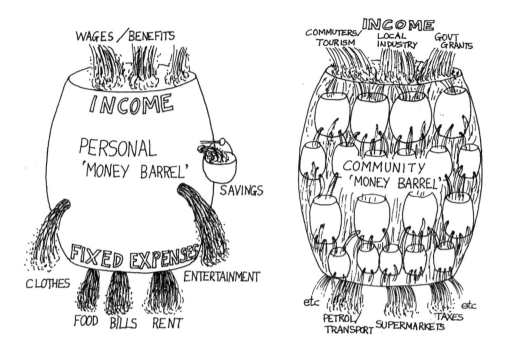

Describe a trade taking place using LETS and show that the seller's account goes up, and the buyer's goes down. Draw a diagram - see below - to illustrate this. I usually use the example of something being repaired, pointing out that there may be cash costs involved, but they get settled in the normal way.

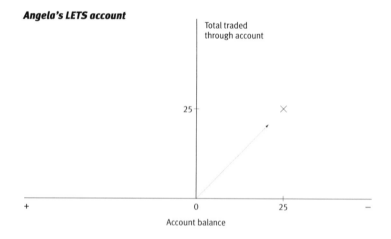

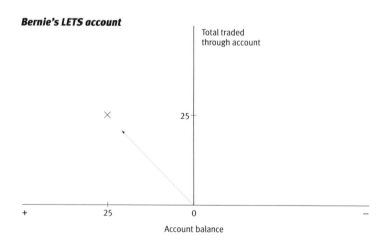

The first trade: Angela pays Bernie £25 for motorbike repair

Usually at this point people either get enthusiastic about the concept or start looking for ways that it won't work. So I just take questions for the next 15 minutes or so. I advise people to use LETS only where both parties can see a win-win outcome.

I tell the story of how LETS came about: the commercial barter networks of North America. I quote Michael Linton on how all the pieces are there; it's how you put them together that counts. I show how the LETS design respects the individual while encouraging mutuality and community. Make the link to permaculture ethics.

I stress the difficulties involved in replicating paper currencies on a local basis. Contrast this with Luis Lopezllera's elegant Tlaloc design which implements LETS in the two-thirds of the world where computer technology is not available.

If there's time and interest, I will mention organisation, illustrating the group-of-three approach to organising within a community (Ref 2), touching on stewardship of individual systems, accounting services and registries, publishing of community information (directories), and publicity and development organisation, including work-netting and the use of trust structures. If there is a general awareness of LETS I will say something about the LETSystem and design by limiting factors – 'the site designs itself'.

Design exercises

Choosing a unit of exchange

I ask: what context? who for? eg LETS for child care measures in hours, whereas a LETSystem measures in credits equivalent to pounds sterling to integrate with mainstream, business taxation etc.

Skills play

Everyone writes down at least three things they can offer to other people - they could be skills, surplus goods etc; also at least three things they would like other people to do for them or provide to them. Then they circulate and see how many trades are possible.

- Everybody attending the session has something to offer.
- Other people may not pay them cash, but they can pay them in LETS.

Points to note

- The measure does not affect value, but using LETS can allow us to value each other more fully.
- The system always balances at zero. A negative balance is a commitment, not a debt.
- The account record follows the trade – other way about in the conventional world!
- The credits in the system are personally issued and the commitments are personally guaranteed.
- No interest is charged.
- The trading community can regulate itself by peer pressure, a self-regulating system.
- People need confidence in the system. Confidence in each other can build more slowly.

Handouts

- Contact addresses (below).
- Frequently asked questions about LETS.
- Copies of bullet points/overheads.

Living examples

Hundreds! Give examples of different applications, eg neighbourhood exchange, community of interest (Permaculture Exchange), voluntary sector (Consensus Building LETS), small business (Manchester and Liverpool), commercial Business Barter Networks etc. Try to build up a good selection of LETS directories to pass around.

Agenda 21 link

Chapter 3. Additionally, throughout the Agenda 21 document, reference is made to the use of 'innovative funding mechanisms'. Could LETSystems be one such mechanism?

Bibliography

1 For more information on the work of Luis Lopezllera and the Tlaloc currency system either:
    Email: mailbase@mailbase.ac.uk with the words 'subscribe econ-lets' or
    Contact Angus Soutar at the address for the LETSystem Trust.
2 Permaculture: A Designers' Manual. Chapter 14

Further research

LETSWork, Lang, P, Grover Books, 1994
Short Circuit
Bringing the Economy Home from the Market, Dobson, RVG, Explains the importance of LETS and how it works
LETS act locally: the growth of local exchange trading systems, Croall, J, Calouste Gulbenkian Foundation, 1997
LETS on Low Income, NEF, 1997
An Introduction to LETSystems, How, F, Letsystem Trust, 1997

LETSLink Scotland – LETS networking in Scotland 31, Banavie Road, Glasgow, G11 5AW.
Tel: (0141) 339 3064

LETSConnect – networking and newsletters for LETS groups. John & Mandy Winkworth, 12 Leasowe
Green, Lightmoor, Telford TF4 3QX. Tel: (01952) 590687

The National Network of LETSystems Registries – details of work-net for general development,
information about legal issues, support of LETS registry services, plus specialist training contacts.
Rob Squires, 34 Devon Street, St Helens, WA10 4HT. Tel: (01744) 612778

LETS Solutions – consultancy and training. 7 Park Street, Worcester, WR5 1AA. Tel: (01905) 352848

The LETSystem Trust – beneficiary support group contacts, specialist training contacts. Angus Soutar,
Ingle Dene, Low Hill, Bury Fold Lane, Darwen, Lancs, BB3 2QG. Tel: (01254) 771555

A starting point for LETS developers: http://www.u-net.com/gmlets/

Land Access

Simon Pratt

Why I include this session

It's difficult to discuss strategies for growing food, fuel etc without some consideration of access to the basic resource. There may be some land access strategies students are not familiar with.

Objective

To discover the possibilities of accessing land in Britain without having to own it.

Learning outcomes

By the end of this session students will be able to:
· Explain how to access land without ownership.
· Identify solutions for access to land for themselves.

Context

Any time. Can be towards the end of course.

Duration

30 minutes

How I teach this session

Small group discussion
15 minutes
Brainstorm the range of possibilities available without having to own a piece of land.

Whole group discussion
15 minutes
Look at one or two of the options developed by the small groups, according to interests of the group as a whole.

Summary
15 minutes
This will include any of the following points, which were not brought up in the previous discussion:
· Historical perspective: Right of use rather than right of ownership (Refs 1, p545, 2 & 3). Tribal groups throughout the world asserting land rights.
· A wide range of options available (Ref 1, p547):
  · allotments or community gardens.
  · city farms.
  · share cropping.
  · community supported agriculture/box schemes (Ref 4).
  · community land trusts.
  · housing co-ops.

- Land Access Office: Lists of people needing land to grow food matched with people offering vacant land – could be older people or absentee landlords (Ref 1, p547).
- Guerrilla gardening: many people actively planting seeds and returning months later to harvest.
- Convert lawns! Find people who will let you use their garden.
- Land Registry: there are many empty buildings around. The Land Registry can tell you whether they have owners or not (Ref 5).

Further activities

- Students can identify opportunities for land access in the venue area on residential courses, or for their own places on non-residential courses.
- Site visits to successful examples, eg city farms or community gardens.

Bibliography

1 Permaculture: A Designers' Manual

2 Land for the People, Girardet, H (Ed), Crescent Books, 1976

3 This Land is Our Land, Shoard, M, Paladin, 1987

4 Local Harvest, Delicious ways to save the planet

5 Land Registry Office, Lincoln's Inn Fields, London, WC2A 3PH. Tel: (020) 7917 8888

Further research

National Federation of City Farms, The Green House, Hereford Street, Bedminster, Bristol, BS3 4NA. Tel: (0117) 923 1800

National Society for Allotment Gardeners, Hunters Road, Corby, Northants, NN17 1JE. Tel: (01536) 66576

Soil Association, Bristol House, 40-56 Victoria Street, Bristol, BS1 6BY. Tel: (0117) 929 0661.

Email: soilassoc@gn.apc.org. Publish a number of directories of organic producers, including Local Food Links which contains some inspiring examples

Official Systems: How They Work and How to Find Out

Judith Hanna

Why I include this session

The problem is the solution: unless you understand how to work with the system, it creates problems, but there are ways to make it work for us.

Objective

To give students the confidence and basic knowledge to engage with local councils and national government policymakers, and understand when it is useful to do so.

Learning outcomes

By the end of this session students will be:

- Unintimidated by jargon and confident to challenge it.
- Understand how to plan a lobbying campaign.
- Know why and how to identify and enlist support from other networks and organisations.

Context

Should follow Local Agenda 21 and precede Transport and Participative Community Planning.

Duration

Parts 1, 2 and 3: 70 minutes
Parts 1 and 4: 65 minutes
All parts: 100 minutes

How I teach this session

1 What official systems?

a) Brainstorm – record as mind-map on flipchart

15 minutes

- What sort of official systems are relevant to permaculture design?
  Should cover:
  - central Government: lobbying MPs and officials to change policies or decisions, getting information.
  - local councils: lobbying councillors, liaising with officers and planning departments.
  - media work, including direct action if lobbying fails.
  - financial systems.
  - international – UN etc – optional.
- How can we work with them?

b) Discussion

20 minutes

- What experiences have students had that they can share with the class?
    - In pairs, three minutes each, about an experience they have had of dealing with official systems. What worked and what they learned from it.
    - Examples: environmental or other campaign, dealing with benefit, tax or other rules, planning permission or setting up a business.
    - Feedback to class. Any main unifying themes?

2 Jargon demolition

20 minutes

Mystifying jargon is a major weapon that establishments and experts use to intimidate ordinary people and to keep them from interfering or taking control away from them. Officials also use jargon to disguise the fact that they don't understand something. Challenging jargon is an important tool in community-level design and planning work.

- Either, in an earlier session, ask students to bring in samples of baffling bureaucratic prose, or bring in the teacher's own collection.
- Student pairs study a sample and translate to plain common sense. Give a prize or acclaim for the best or wittiest translation.

3 Role-playing

15 minutes

- In fours, each with a scenario which can be taken from 1b) above.
- Two in each group are the official side, the other two trying to influence or understand them.
- Five minutes role play, then change roles.
- Alternative scenario: two are reporters interviewing the other two.

4 Campaign design

(also relevant to 'Participative Community Planning', see pages 124-126)

30 minutes minimum

- Especially useful where students are working on a design project, or if they are involved in a campaign or lobbying. Take them through the steps of planning a campaign. These are:
- Identify objectives: what do you need to achieve/change?
- Identify/research official objectives: relevant policies, guidance, law? What in their policies supports your objectives? What needs to be changed?
- Who you need to influence?
- Allies: how do you reach them and enrol their support? what support might they give? useful networks, media.
- Opposition: how do you convert or neutralise it?
- Other resources you can tap: what resources are needed? How can you secure them?
- Time constraints and targets: schedule to achieve them.
- So what's your action plan?

This exercise can be taken away as individual homework, to be fed back or displayed at a later session. Or it can be a half-hour small group exercise, fed back as the culmination of the session.

○   **Further research**

Lobbying, an Insiders Guide, Dubs, A, Pluto Press, 1989

CPRE publishes a series of campaigners' guides on a range of planning issues, eg roads, minerals extraction. Local Council committee papers and reports should be available for reference in libraries, which should also maintain a list of local voluntary organisations, and may keep the Hansard parliamentary reports.

Local Agenda 21 networks should have most of the useful local contacts, including the local authority and any local Council for Voluntary Service or volunteer bureau. Directories of useful national organisations, eg those published by NCVO and the Environment Council, should also be available in public libraries.

Outstanding examples of confusing jargon can be sent to the Plain English Campaign, PO Box 3, New Mills, Stockport SK12 4QP.

Start at the Front Door? How I Teach the Application of Permaculture

Angus Soutar

Why I include this session

Since permaculture is primarily about systems and design, you can do it anywhere. For me, the most powerful way to communicate this message is to make the maximum use of the environment in which the course is taking place. Similarly, we can run the Design Course almost anywhere. Everything, both in the training location and immediately outside it, is a potential learning resource.

Students can live in a surprising variety of locations, and there are often individuals who are on the move and planning to live in an even more surprising variety of locations. So I also find it useful to enhance the understanding of permaculture by examining its application to environments other than the course setting.

I want students to be empowered to act, rather than disempowered because they think they need access to land or expensive technical resources.

I also like to do some work which encourages co-operative efforts to start tackling the wider issues that tend to overwhelm people as individuals.

Finally, I feel a growing need to promote a better understanding between urban and rural communities, while giving information, and hopefully some inspiration, about how beneficial links can be built between the two.

Objective

To demonstrate that permaculture can be applied anywhere, so that anyone (everyone?) can contribute to achieving a sustainable way of life.

Learning outcomes

By the end of the session, students will be able to:
· Participate in a group problem solving process.
· Identify improvements that can be made to a neighbourhood to make life more sustainable.
· Define 'rolling permaculture'
· Discuss the differences and similarities in applying permaculture in two different environments
  (eg rural and inner-city, both in a temperate zone).

Context

Some of the key points are covered in other course topics such as Design Exercises and the sessions on Local Food Links, Land Access or The Urban Environment. Others can arise during and after site visits. This session pulls key points together and is an opportunity to fill some of the remaining gaps.

I like to do the visioning exercise early on in the course. As well as generating ideas for the Next Steps process for individuals, it gives me pointers towards suitable design exercises, which is important on courses where land access is a problem. The rest of the material can be used in the second half of the course.

○ **Duration**

Part 1: Two sessions of 45 minutes

Part 2: 45 minutes

Part 3: 10 minutes minimum

○ **How I teach this session**

Part 1: Local setting

Visioning exercise (early in the course)

45 minutes

I find this particularly useful when some of the participants are local residents. I ask the group "What would make this a better place to live in?". With the group, organise the results into a diagram on the wall. Keep the diagram available on the wall throughout the rest of the course. The participants' replies will tend to be inversions of the major current problems in the area. I try not to articulate the problems at this stage, staying positive.

Problem tree

35 minutes

Some time after the visioning exercise, we do a 'problem tree'. I draw a tree on a very big piece of paper, with both roots and branches. (See diagram below.) Evident problems are placed at the top of the tree. Through group discussion we establish what the root causes of these problems might be. These are placed at the bottom. The problems can be written on post-it notes, so that they can be moved around easily if people change their minds about whether they are symptoms or causes. This exercise helps people to focus on some of the more persistent local problems. I have based these exercises on Joanne Tippett's work in Africa. They work particularly well in urban areas in the UK.

We then have some discussion about whether these results are representative of the local community in general. Then I make some suggestions as to the applicability of permaculture in addressing the problems.

(Note: I photograph and keep the results of the above sessions. I am working with Rob Squires to compare and contrast these photo-reports. We invite others to join in this research with us.)

Guided discussion

10 minutes

These results are still too wide ranging for us to use in a practical way. Now I write up a permaculture principle: Start at the Back Door. We have a group discussion about the meaning of this. I summarise by saying that starting at the back door means starting where I am. And if I start from anywhere else then I am deluding myself.

What do I do at the door? I can choose to start with the door itself – perhaps looking at the potential to cut domestic energy loss. Maybe I see an opportunity to grow something next to the doorstep. Or maybe I decide to walk down the street and find someone I can work with to improve sustainability in the neighbourhood. This is even more important in an urban context, where, in many cases, dwellings don't have back doors – you're out on the street before you find a suitable place to site a herb planter. So be prepared to start at the front door!

I bring in the importance of taking small achievable steps, and how it often becomes easier when you can find a few friends to work with. Rolling Permaculture can also be explained here (a process of incremental design).

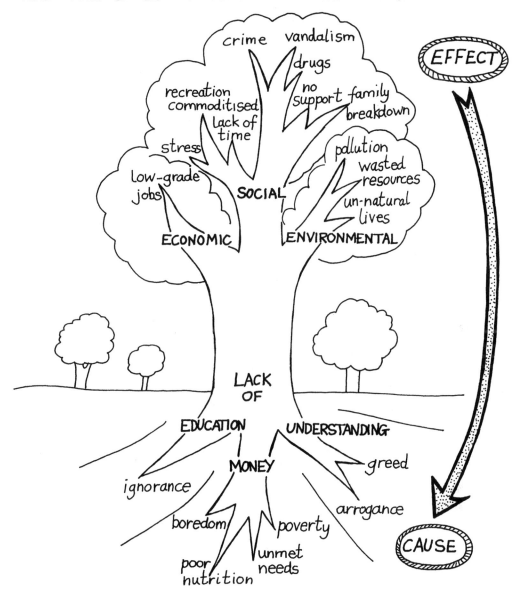

What are the main problems in your life, and the lives of people around you?

Part 2: Comparison of two or more environments

Brainstorm

10 minutes

List different types of living environment. In the UK, usually the group will come up with something like: Inner City/Suburbs and 'Consolidated' Villages/Urban Fringe/Rural Village/Deep Rural.

(More outward-looking participants may be more interested in global climate zones: tropical, arid, Mediterranean etc, in which case I go and get either a video (Ref 1) or a visiting teacher from overseas, whichever is closer to hand.)

List characteristics

10 minutes

I choose some examples from the list, favouring ones where participants have lived. We then list some characteristics of these areas. Either a SWOT (Strengths, Weaknesses, Opportunities, Threats) or PMI does the trick.

Discussion

10 minutes

Through more discussion, we develop some general design strategies, eg suburban and urban fringe areas could offer more short-term opportunities for sustainable food production than the agricultural wastelands of East Anglia. Setting up a recycling project may be a better use of our resources in the inner city than trying to grow food. I emphasise the principle of Least Effort For Greatest Effect. The results will probably be tentative, because we haven't got enough information. This is where the need for further research comes up.

Variation

15 minutes

Compare and contrast two very different environments, especially inner-city and rural. What beneficial links can be set in place between communities in the two areas? For example, young people in rural areas may want a safe space to stay in the city; city people may want to get food from the rural areas, go WWOOFing etc.

Part 3: Further inspiration

10 minutes minimum

Show some video from the Global Gardener series (Ref 1).

Further activities

For further consideration:
·   What do urban people want from towns/the countryside?
·   What do country people want from the towns/the countryside?

This can lead to a longer-term visioning exercise – 'Two scenarios' – the sub-urbanisation of the countryside versus the re-greening of the city (urban orchards and market gardens).

Bibliography

1 Global Gardener

Further research

'Urban oases – greening the urban desert', Milroy, A. ,Article in Town and Country Planning, October 1995

Convenor's Guide

Convenor's Guide

George Sobol and Patsy Garrard

The aim of this section is to provide a pattern and checklist of various aspects of course convening, so that you will be able to plan and run a successful one!

This outline represents work in progress.

· Each course will be specific to its territory.
· Each course adds to our knowledge about convening successful courses.
· Each course is a design – apply permaculture ethics and principles.
· Let your course model the ethics and principles.

1 Well in advance of the course

1.1 Who convenes?

· An individual:
  · who wants to attend a course.
  · who wants to apprentice as a teacher or convenor.
· A group:
  · geographical community.
  · intentional community.
· An organisation:
  · with shared aims and ethics.
· A Further and Adult Education Provider as part of a programme of courses:
  · Workers' Educational Associations.
  · Colleges of Further Education.
  · Community Colleges.

1.2 Course format

Here are some possible formats for the Design Course:
· Introductory weekend – can be free standing or first part of a Design Course.
· A series of weekends.
· A two-week residential.
· An evening class over two terms with a practical weekend each term.
· A correspondence course/distance learning – an opportunity for development.
· As part of another, longer course.
· As a combination of weekends and evening classes.

The following PMIs on course formats (Plus, Minus and Interesting points) were generated by Simon Pratt, with a few points added by us:

Two week residential:

Plus: immersion in permaculture.

stable group.

live and breathe permaculture.

removed from day to day life.

eating together (very important).

Minus: intensive.

little time for reflection.

information overload.

cost.

Interesting: group dynamic.

relationship with site.

Series of weekends:

Plus: time for reflection, relate to own situation.

inexpensive.

can fit around other commitments.

eating together.

Minus: outside distractions.

lack of social time.

group may change.

extra travel (time/cost/energy).

loading/unloading materials.

Interesting: homework, reading.

seed for local group/project.

Weekends plus one week:

Plus: variation in experience.

immersion and reflection.

flexible.

can fit around other commitments.

eating together.

Minus: cost.

difficult to maintain intensity of residential week.

Interesting: homework, reading.

Evenings:

Plus: new audiences.

can fit around other commitments.

time for reflection, relate to own situation.

inexpensive (or free).

Minus: restricted access to sites.

limited time each session.

group may change.

people tired after a days work.

extra travel (time/cost/energy).

loading/unloading materials.

Interesting: homework, reading.

seed for local group/project.

In 1997 we succeeded in gaining accreditation for the Design Course with the Open College Network (OCN) via the Workers' Educational Association (WEA). If you wish to run the OCN-accredited course please see the section on permaculture courses and the WEA (see page 366-367).

1.3 Venue

Evening classes are usually run using existing Further and Adult Education venues:

Plus: AVA (audio visual aids) often available.

'Edge' with other courses/the community.

Minus: people tired (end of day/travel effort).

Inflexible rooms (furniture to move/strip lights).

Interesting: develop mobile reference library.

go to pub afterwards for social time.

The following applies to venues for weekend and residential courses:

Courses are often run at venues that do not normally run or host courses. Even if they do, it is important to visit before going ahead with the planning of the course. Don't forget that all comers should feel comfortable and at home on a course.

- Visit with as many of the course convening team as possible (see 2.1).

- Check the following:
  - ☐ Teaching room:
    - ☐ needs of the teachers and the participants.
    - ☐ enough room for 20-24 people.
    - ☐ variety of seating options from chairs and tables to floor cushions.
    - ☐ warm in winter, cool in summer.
    - ☐ power supply for audio visual aids (AVAs).
    - ☐ can it be blacked out for slide shows?
    - ☐ other rooms for small group work.
  - ☐ Kitchen and a room to share meals:
    - ☐ is it equipped to cater for a large group?
  - ☐ A quiet room for library, quiet space and sick bay, separate if possible.
  - ☐ Accommodation for teachers and, if offered, for participants:
    - ☐ bedrooms, camping and local B&B options.
  - ☐ Loos and washing:
    - ☐ are there enough loos?
    - ☐ will the water supply cope with a large group for two weeks?
    - ☐ will the sewage system cope with a large group for two weeks?
  - ☐ Health and safety, public liability:
    - ☐ Disabled access.
    - ☐ Parking and public transport options.
    - ☐ Childcare – needs a room or area specifically for this plus qualified people to run it.
    - ☐ Opportunity for practicals on the site.
    - ☐ Opportunity for walks and observation exercises.
    - ☐ Is this a venue where a design will be adopted?

You can now agree dates.

1.4 Costings

May include the following:

- ☐ Accommodation.
- ☐ Catering.
  On weekend courses it works exceptionally well to ask people to bring food to share. It results in feasts of enormous variety!
- ☐ Childcare.
- ☐ Contribution to a bursary fund.
- ☐ Convenors' fees and expenses – photocopying, postage, phonecalls, etc.
- ☐ Permaculture Association membership for one year.
- ☐ Publicity.
- ☐ Tutors' fees and expenses (including visiting tutors).
- ☐ Venue costs.
- ☐ Visits and transport.

Do you have a strategy for bursaries or sliding scale of course fees?

An upward, as well as downward sliding scale of course fees gives participants in well-paid occupations the opportunity to help those on low or no incomes.

For further information on costings see section on permaculture courses and the WEA (see page 365-366).

1.5 Publicity

You are now ready to put out your course publicity. This will include:

- [ ] Venue.
- [ ] Dates.
- [ ] Convenor's name, contact address, phone number, fax and email details.
- [ ] Brief introduction to permaculture.
- [ ] Course outline – content, practicals, participatory, etc.
- [ ] Tutors' names and backgrounds.
- [ ] Availability of apprenticeships for trainee teachers.
- [ ] A booking form to include:
  - [ ] participant's name, address and phone number/fax/email.
    say Please print or type – and allow plenty of space so that it comes back clear and legible.
  - [ ] fee options.
  - [ ] deposit/full fee.
  - [ ] policy on refunds if participant cannot attend or if the course does not run.
  - [ ] closing date for bookings.
  - [ ] special needs – dietary/health/access.
  - [ ] childcare.
  - [ ] can they offer a lift?
  - [ ] do they need a lift?
  - [ ] contact in case of illness.

  NB: Do not print the booking form on the back of essential information you want participants to keep!

To publicise the course you can use:

- [ ] Editorials in local papers and magazines
- [ ] If you are working with a Further and Adult Education Provider, listing in their prospectus
- [ ] Free listings in permaculture publications and web pages and, if there is one, your local Green Events
- [ ] Posters and fliers – it's easier to place A5 posters on notice boards than bigger ones!
- [ ] Selected mailing to previous enquiries, past participants on your courses and allied organisations.
- [ ] Word of mouth – use your allies.
- [ ] Selective advertising.

2 Pre-course admin

2.1 Course team

Don't try to do it all on your own!

Identify your course convening team. Individuals may have more than one role but ensure back-ups for each role. These will include:

- Apprentices:
  - act as back-up for the other team members.

- Caterers:
  - probably the most important role on a residential course.
  - vegetarian food – easier for storing and washing up, and saves catering separately for veggies and carnivores (organic meat option if demand once course has started).
  - local, organic produce where possible.
  - agree a budget.

- Convenors:
  - putting it all together.
  - responding to bookings and enquiries (see 2.2).
  - finances.

- Tutors:
  - timetable.
  - handouts.
  - practicals in consultation with site crew.
  - visits to local sites of interest.
  - guest speakers.
  - teacher training for apprentices.

- Site crew (to include somebody from the host venue):
  - materials and tools for practicals.
  - needs identified during visit to venue.

Clarify roles and mutual and personal expectations (including financial) – write them down!

2.2 Paperwork

- Response to bookings:
  - receipt.
  - 'thank you for booking' letter.

- Along with this you will send:
  - joining instructions.
  - map.
  - public transport options.
  - arrivals/start time.
  - day off (on residentials).
  - finish time/departures.
  - what to bring:

- · food to share (on weekend courses).
- · mug, bowl, plate, cutlery.
- · notetaking and drawing materials.
- · clipboard/drawing board.
- · compass.
- · pocket calculator.
- · boots and fine/foul weather gear for practicals.
- · indoor shoes or slippers.
- · books and magazines for library.
- · seeds and plant material to share.
- · musical instruments.
- · -information on own site or project.
- · Provisional timetable.
- · Course information (see Appendix for example).

2.3 Apprentices

- Actively encourage apprentices to participate on weekend and residential courses.

- Apprentices can be invited via the course publicity or by direct invitation of previous course participants.

- During this pre-course stage the tutors can 'take bids' from apprentices – what sessions would they like to teach? Alternatively, where the tutors know apprentices from a previous course, they can suggest sessions for the apprentices to prepare.
  Three ways of deciding what to teach:
  - · Teach your passion.
  - · Teach what you have direct experience of.
  - · Teach what you need to learn.

- Send copies of own session notes as guidance to apprentices if appropriate.

- Joining instructions for apprentices. Send same as for participants but additional notes to include:
  - · proposed sessions.
  - · notes on session preparation and presentation.
  - · arrivals and departures.
    - · day before on weekend courses.
    - · several days before residential courses to give time for practice sessions with feedback as well as helping in the setting up of the course.

3 Immediately before the course

3.1 Whole course team

- Meet up immediately before the course – the day before a weekend and three to four days before a full residential course. Establish format for meetings.
  - Check in – how's everyone feeling, eg three minutes each.
  - Set timing for the meeting.
  - Agenda items – what do we need to discuss?
  - Allocate available time.
  - Everybody heard.
  - Check out.

- Course team:
  - Review roles, needs and mutual expectations.
  - Identify a 'flack-catcher' or 'switchboard operator' for the course – their role is to take comments or suggestions from the course participants relating to the course, venue, catering, etc. and relay them to the appropriate course team.
  - Address any anxieties.
  - What still needs to be done?

3.2 Finalise arrangements

- Venue:
  - Reception area (on full residential and first weekend of series of weekends).
    This is the first impression that participants get when they arrive.
  - Map of the site and good signing so participants can find their way around.
    You may need to put up signs in the surrounding area if the site is particularly hidden!
  - Place for wet clothes and muddy boots.
  - Teaching space.
    Decorate with rugs, cushions, fabrics, plants – it will be your home for the next two weeks.
  - Kitchen and eating room.
    Food hygiene – space for storage of prepared food brought to weekend courses, dog and cat free
    Waste separation system – compost, recycling bins, etc
  - Quiet room, library, sick bay and first aid kit.
  - Loos and hand washing.
  - Accommodation.
  - Practicals – tools and materials.
  - Visits – transport and timing.

- Finalise arrangements for the day off (on residential courses):
  - Arrange an optional, purely recreational trip to the seaside, the woods or some other local attraction.
  - This is also a day off for the catering team.

- Agree daily pattern for the course.

3.3 Apprentice support

- Teaching team meets.
- Use same meeting model as for convening team (3.1).
- Apprentices shadow key course convening team roles to get experience of running courses.
- Timetable practice teaching sessions before the start of the course.
- Update timetable, finalise practicals and visits.
- Make sure all members of the teaching team have all the support they need for preparing and delivering sessions.

4 During and after the course

4.1 First day

Most of the first morning or the first evening of a residential course is taken up by introductions:

- Registration as participants arrive.

- Collect any outstanding fees.

- Welcome to participants.

- Introduction to the course team.

- Introduction to the venue:
  - House rules, eg no smoking, no shoes, etc.
  - Domestic arrangements.
    - Catering and clearing up teams on weekends.
    - Housekeeping teams on residentials.

- Introduction to the course:
  - The course as a qualification – national and international context.
  - Participants expected to take notes (session on mind-mapping at start of course) to produce personal learning and evaluative journal. Of value to participants in the future and as evidence of course completion.
  - Participants working in groups on a design project at the end of the course.
  - Expect attendance – if you have to miss a session find an ally to help you catch up.
  - If you do not understand then ask.
  - Time for questions and discussion during each session.

- Course culture:
  - Sessions will start and finish on time (volunteer timekeeper/sheepdog?).
  - Rest if you are tired – quiet room.
  - Library and books for sale.
  - Can you offer photocopying?
  - Offers and requests posters for everyone to use during the course.
  - 'Cheating' encouraged – co-operation not competition.
  - Opportunity for feedback from participants half-way through the course as well as at the end – course convening team responsive to needs and comments of participants – student centred learning.
  - Day off half way through residential course.
  - Cabaret on last night of the course.

- Brief introduction to the participants (more detailed introductions next day):
  - Name.
  - Where do you come from, where is home?
  - What are your expectations of the course?

- If people arrive late, make sure they have an opportunity to introduce themselves.

- Introduction to permaculture!

4.2 During the course

Daily routine:

- Course convening team meets early in the morning:
    - Check in – how's everyone feeling, eg three minutes each.
    - Set timing for the meeting.
    - Agenda items – includes items for the opening session.
    - Allocate available time.
    - Everybody heard.
    - Check out.

- Opening session:
    - Three minutes each way to help participants 'land'. One person introduces themselves and then speaks while the other listens. After three minutes they change over. This also helps develop listening skills.
    - Whole group in a circle so that everyone can see everybody else's face.
    - If it seems appropriate, a minute of silence.
    - Round of names – name and… eg nearest river, nearest hill or mountain, favourite tree, word to describe how you are feeling this morning, etc. Helps the group learn each other's names.
    - Some sort of physical exercise to get the blood oxygenated. Short yoga sequence or a quick game.
    - Whatever a member of the convening group or course participants can lead.
    - Domestic announcements.
    - Course announcements.
    - Participants announcements.
    - Run through the day's timetable/any adjustments.

- Teaching sessions:
    - 45 minutes maximum or double session with break.
    - Time for questions, discussion, groupwork, etc.
    - Practicals or visits in the 'graveyard slot' after lunch.

- Visits:
    - Visits to local sites that demonstrate permaculture in practice or techniques described in the course (usually one afternoon per week on a residential course).

- Breaks:
    - Long breaks – half an hour morning and afternoon, one and a half to two hours for lunch and supper. The breaks are the rich 'edges' of a course where productive contacts and connections are made (not to mention a 15-minute siesta after lunch). Also depends on size of site. Camping needs the extra time for 'housekeeping'.

- Course convening team meets at the end of the afternoon sessions:
    - Check in – how's everyone feeling (eg three minutes each).
    - Set timing for the meeting.
    - Agenda items – including feedback on the day.
    - Allocate available time.
    - Everybody heard.
    - Check out.

- Evenings:
  - Optional programme – cover design course material during the day.
  - Courses are very intensive – allow time for rest and recreation.
  - Give participants opportunity to give presentations.
  - Videos and slide shows of associated interest.
  - Teachers meet with apprentices – can use same format as course convening team meeting. Feedback on the day's teaching. Timetable adjustments and session planning for the next day. Try not to disappear for too long.

Halfway through the course:

- Halfway evaluation of the course so far. Start with five minutes 'think and listen'. Same as three minutes each way of the opening but participants use the time as 'thinking aloud' time. Allows for concise reporting back on:
  - One thing I particularly liked.
  - One thing I would have done differently.
  - One thing I am looking forward to.

  Divide available time equally between everybody, including the course convening team, for feedback.

  Stress the I-factor for the 'done-differently'. The aim is to get constructive suggestions rather than negative 'I didn't like...' comments.

- Day off:
  - Optional day out for the group (on residential courses) – purely recreational.
  - Others stay on site or go off separately.

  The day off is very important on residential courses. Don't forget you are asking people to assimilate a lot of information and some challenging ideas in a short period of time. A break is essential. Tests have shown that 'wobblies' happen at the end of the first week!

4.3 Endings

- Cabaret and party:
  - Last evening of the course.
  - Encourage all participants to prepare a turn, including the course convening team! You can judge the success of a course by the cabaret night.
  - Cabaret compres, volunteered from among the participants, plan the evening and present the show.
  - Participants can invite friends and family, where local.

- Last day
  - Where Do We Go From Here? – exit strategies.
  - Make sure everyone has an address list of all participants.
  - Collect in Permaculture Association Membership Forms where this is included in the cost of the course. (This is strongly encouraged.).
  - Paperwork (if working with educational organisation such as WEA).
  - Verbal evaluation of the course.
    - Start with five minutes each way 'think and listen' (see halfway feedback).
    - One thing I particularly liked.
    - One thing I would have done differently.

- · My long-term goal.
- · My next achievable step.
- · One thing I am looking forward to.
- · Presentation of course completion certificates
  Arrange something special to mark the end of the course. By the end of a course there is a closely knit group and folk are sorry to be parting.
- · Farewells.
- · Clear up.
- · Lifts to railway and bus stations.

4.4 After the course

- · Course convening team meets after the course:
  - · Evaluation of the course in general.

- · Teaching team meets:
  - · Evaluation of the teaching.

- · Convenor:
  - · Send list of participants to the Association office along with membership forms and subscriptions, where applicable.
  - · Update own database.

- · Course convening team take a well-deserved rest!

Acknowledgements

This overview of course convening is the product of work we have done over the last nine years with various course convening teams.

Particular acknowledgments are due to Liz Roberts and Andy Langford, George's first co-convenors on the Design Course in Dartington in 1989; to Lea Harrison for her Permaculture Teacher Training Course at Middle Wood in 1990; to Simon Pratt, Patsy's first co-teacher at Redfield in 1991; and to Karol, Anna and Patricia Koncko, who have worked with us since 1994 developing courses in the Czech Republic and Slovakia.

Appendices

About the Permaculture Association

The Permaculture Association is a national educational charity promoting permaculture and developing the awareness and practical skills of its members. It acts as a vehicle for connecting people, ideas and resources to projects in Britain and throughout the world.

Some history...

One of the founder members of the Permaculture Association writes, "My first experience of Permaculture was a four-day introduction in the South of England, in the Autumn of 1982. Soon afterwards I was amongst a group of about six people at a Design Course in the tiny Lake District village of Blencarn. This was given by Max Lindegger from Permaculture Nambour. On the final weekend we climaxed in what must have been the first Permaculture Britain convergence. At this gathering, Max invited us to form ourselves into the first British Permaculture Association, somewhat against our own inclinations I seem to remember. A small sub-group of us agreed to edit the Newsletter, and I remained a part of this for about three years.

"During this time the Permaculture people, numbering something like 10 in total, seemed to be little more than a band of off-the-wall starry-eyed lunatics, who wondered if the world might conceivably 'catch up with us' one day, or perhaps would simply never know we ever existed. We felt that our mission was to keep a spark of Permaculture thinking alive, in case, or until such time as, more interest was able to build up. On the practical level, there were exciting local initiatives struggling to happen in various places, but sadly our tiny Permaculture Association was not able to provide real support for any of these.

"The turning point seemed to be in the Summer of 1985, after a successful lecture tour by Sego Jackson from the Permaculture Institute of North America. The lecture tour culminated in a gathering of about 40 people and a fresh leadership initiative from a group of the new design graduates who were better placed, and had the much-needed fresh energy and commitment, to carry the movement forward to another stage."
Written by Mike Roth

And today...

The Association supports thousands of individuals, projects and groups working with permaculture in Britain by:
- Running a membership scheme to inform, support and network activists, projects and groups.
- Supporting the development of permaculture education and training.
- Organising national and local events.
- Networking nationally and internationally.
- Providing information to its members, the media and the wider public.
- Seeking funding to research and develop permaculture projects.
- Working to make permaculture accessible to all people in Britain.

The charity is run by a Council of Management, who meet quarterly. This Council is elected each year at the AGM. Any member can stand for election and all are entitled to vote. The AGM itself is a small part of the annual Permaculture Conference and Convergence and details about this are published in the Association's newsletter Permaculture Works!

○ *Staff and volunteers*

The central office has a small team of dedicated staff, who deal with enquiries, run the membership scheme and put together the newsletter. The Association is also fortunate to have a good number of volunteers who work on a range of different projects. If you have time to spare, why not contact the office for current opportunities.

○ *Membership*

There are three levels of membership: associate, graduate and group. Contact the office for a joining form and details of the particular benefits of each category.

○ *Information available from the office:*

· Listings of permaculture Introductory, Design, specialist and other environmental courses.
· Local groups lists. An updated list is sent to all members annually, or on request.
· General information about the work of the Association in Britain and what's happening globally.
· Sites to visit.
· Information sheets covering topics such as, transport, Local Agenda 21, LETSystems and fundraising.
· Permaculture videos, which are available for hire.
· The Designers Register, which is a listing of active designers and teachers.

The office maintains a resource bank of information, magazines, library video tapes, teaching material and local group newsletters which is available for your use. We are encouraging local groups to establish their own resource banks. If you want to do this then please call the Association for a listing of recommended books. We can also help you to track down any specific information you require.

○ *Communications*

· Permaculture Works, our quarterly newsletter.
· An ever evolving website.
· Our exhibition and literature is available for shows, fairs and festivals.
· Our office is open from 10am to 2pm on Tuesdays, Wednesdays and Thursdays for telephone enquiries.
· The Annual Convergence is the largest permaculture gathering in Britain and serves to bring us all together to share news and ideas and give us focus for the coming year. A wide variety of workshops, discussions, lectures and social events are held throughout the event.

○ *Introductory courses*

The Association recommends that individuals who wish to know more about permaculture start by attending one of these two-day courses.

○ *72-hour Permaculture Design Courses*

These are offered as two week intensive courses or in modular form, eg as evening classes, and lead to the Permaculture Design Certificate. The Association promotes the Design Course as a vital stage in developing an understanding of permaculture ethics, principles, design processes and implementation techniques.

○ **Diploma**

Having completed the 72-hour course, graduates may consider working towards accreditation for the Diploma in Applied Permaculture Design. This is a recognition of active work in applying the ethics, principles, design strategies and operational techniques. A diploma holder is qualified to teach permaculture design. See page 351-352 for further details.

○ **Teachers Group**

The Teachers Group meets annually, the day before the Convergence and AGM. Other meetings and activities may also be arranged. It has a wide and growing range of materials which can be made available to new teachers. Contact the office to be added to the teachers group mailing list.

○ **Support for groups and projects**

The Association connects projects and groups of all sizes and types via the Permaculture Project Network (PPN). A directory of PPN members is produced so that members can contact and support each other easily. It promotes good project design which integrates economic, social and environmental elements. It also provides advice, case study material, research and networking with other organisations.

Members of the PPN can apply to use the Association's charitable status to help raise funds for projects. To do this, contact the office and we will send you an application form which is passed on to the Council of Management. If they feel it is a suitable project, which is not likely to harm the good name of the Association, they will give it approval.

○ **Local groups**

These are the main focus of permaculture activity in Britain. They differ widely in nature, objectives and activities. Some simply have monthly meetings where the members exchange information, listen to lectures etc. Other local groups are more like communities with their own food supply systems, LETSystems and projects. Local groups can be group members of the Association. Contact the office for details. Local groups often set up Permaculture courses and the office can provide lists of teachers. The office can supply you with a list of local groups, and if there isn't a group near you then the office can supply a local group start-up sheet and a list of Association members near you. Let us know if we can help. When groups run courses, give talks or put on public displays, then a box of display books is available though co-logic books and Permanent Publications, who can also provide discounts for bulk orders. Leaflets, membership forms and display materials are all available from the office.

○ **Visiting sites**

Most permaculture projects and sites are happy to have visitors and volunteers, and the office can supply information about them. The Permaculture Plot from Permanent Publications lists over 86 places throughout Britain that can be visited. Each entry describes the project, often accompanied by a site plan or other illustrations. Full contact details are given.

The Association can also provide details of projects and groups around the world.

Discounts and services

The Association regularly negotiates discounts for its members including books from co-logic (5%), computer equipment, and ethical green services, eg discounted membership of the Environmental Transport Association. Please contact the office for further details.

To find out more, or become actively involved, contact the Association at:
BCM Permaculture Association, London, WCIN 3XX.
Tel: 0845 4581805
Email: office@permaculture.org.uk
Website: www.permaculture.org.uk

⇨ Working for the Diploma of Applied Permaculture Design with the Permaculture Academy (Britain)

○ *Your Action Learning Pathway to the Diploma from your 72-hour full Design Course*

doing project work

observing effects

thinking about theory

designing next steps

The Academy will support you in designing your own pathway to the Diploma. Your pathway will be unique to you and, because you design it yourself, tailored to suit your circumstances and requirements.

A pathway design includes elements to do with:
· Content – what you work on.
· Process – how you work on it.
· Timing – how you sequence the work.

Some of the elements of content, process and timing are set by the Academy to provide a framework for your pathway design. For example, all pathways to the Diploma have in common a minimum of two years applied, part time or full time, permaculture design work as a requirement.

Activities during the minimum two year period are about learning by doing, and are called Action Learning.

○ **Registration with the Academy**

Note that you must be a paid up member of the Permaculture Association (Britain) before you can register with the Academy.

Registration brings you the benefits of Academy support while travelling your Action Learning Pathway. Registration is a two stage process.

Firstly you order a registration pack. (Note to teachers: the office can supply leaflets and order forms on request.)

Secondly, after you have looked through the registration pack, you prepare your application for registration as an Apprentice Designer. This includes submitting a draft design for your chosen Action Learning Pathway. How to prepare a draft pathway design is explained in the registration pack.

Apprentices and Guides

While you are learning permaculture design skills through your Action Learning Pathway you will be an Apprentice Designer. The Academy will help you to identify and work with a Learning Guide, a person who has experience of Action Learning and can support you in developing your own skills. Your guide will also have a good knowledge of the permaculture network and can direct you to tutors with technical skills appropriate to your needs.

Developing the Academy

The Academy was founded in 1991 and supports Apprentices in designing and following their Diploma pathways. The Academy is developing continually and your feedback, as a participant, will be sought and used to fine tune the support services we offer and to guide us in initiating new services as required.

You can also gain a Diploma through the Permaculture Institute of Australia, PO Box 1 Tyalgum, NSW 2484, Australia. The Institute is currently a 'support free zone' and thus suits go-it-aloners.

Academy Publications

The Academy publishes a range of guidance booklets designed to support you in navigating your Action Learning Pathway. Details of guidance booklets published to date will be sent to you as part of the registration pack.

Action Learning – the core skill in your pathway

Action Learning is a cycle which includes the following elements:
· Doing practical project work.
· Systematically noting your observations about effects.
· Thinking about how your experience affects your understanding of permaculture theory.
· Working out how your conclusions will affect your designs for the next action opportunity.
· More practical project work incorporating your new learning.

The purpose of your Action Learning Pathway is to convert the theory you learnt on the Design Course into theory grounded in your own experience.

Fees

The Permaculture Academy seeks to provide the support you need at prices you can afford. A range of strategies is available:
· Pay as you go – to even out the costs to you.
· Sliding scales – to adjust the fees to match your income and wealth.
· Payment in Stars – the currency of the Permaculture Exchange system.

We operate an open books and clear costings policy for your benefit. Details are in the registration pack.

Your next step

Apply to the Academy at the office address for the appropriate forms. The Permaculture Academy is part of the Permaculture Association (Britain).
Write to: BCM Permaculture Association, London, WCIN 3XX.
Tel: 0845 4581805
Email: office@permaculture.org.uk

⇨ *Teaching Resources*

○ ### *First and foremost – each other!*

The Teachers Group organises an annual gathering of permaculture teachers, and other meetings on an occasional basis. It can also organise training and has copies of most of the handouts that are mentioned in this guide. Contact the Association to be added to the mailing list.

○ ### *Sources of training*

City & Guilds Further & Adult Education Teachers Course – WEA. See previous section.
Participatory Training Techniques – ATB Landbase, contact Graham Bell on tel: (01835) 822122.
Neighbourhood Initiatives Foundation, The Poplars, Lightmoor, Telford, Shropshire, TF4 3QN. Tel: (01952) 590777. Fax: (01952) 591 771. Offer courses in Planning for Real.
HSDI: A Human Scale Development Initiative, 8b Vicars Road, Leeds, LS8 5AS. Tel: (0113) 2400 349.
Training in Thinking Tools, meeting management and more.
Centre for Alternative Technology, Machynlleth, Powys, SY20 9AZ. Tel: (01654) 703743. A wide range of courses including organic systems, water, wind and solar energy. Contact Joan Randle for more details.
Accelerated Learning, 50 Aylesbury Road, Aston Clinton, Bucks, HP22 5AH. Tel: (01296) 631 177. Fax: (01296) 631074.
Apprenticeships – many teachers are willing to provide opportunities for graduates to start teaching. Contact the Association for details of your nearest teachers.
The Institute for Earth Education is a 'Head, heart and hands' approach to helping people develop a sustainable and personally nurturing relationship with the natural world. Complete programmes of education are designed to engage the learners and motivate them with highly participatory activities. These are focused on helping them to develop understanding of, and feelings for the natural world, and to encourage positive changes in their lifestyles: to 'Live more lightly on the planet'. Model programmes are published along with activities and guidance for those wishing to develop their own earth education programme. An international catalogue of resources for teaching, books and items is available from the PO Box address along with a UK price list and order form.
PO Box 91, Tring, Herts, HP23 4RS.
Tel/Answerphone/Fax (01442) 890875.

○ ### *Permaculture teaching venues*

This listing gives details of teaching venues that are also practising permaculture sites. Many of them can be found in The Permaculture Plot. Another useful publication for teachers and facilitators looking for a suitable venue is Places to Be, How, J (Ed), Coherent Visions, 1997. There is an online website at http://ourworld.compuserve.com/coherentvisions/placestb.htm

England
· ***Earthworks Sustainability Centre,*** East Meon, Hants, GU32 1HR.
  Contact John Tugwell on (01730) 823 166
· ***Earthworm Housing Co-op,*** Wheatstone House, Dark Lane, Leintwardine, Craven Arms, Shropshire,

SY7 0LH. Contact Hil or Gaea on tel: (01547) 540461.

· **HDRA,** Ryton-on-Dunsmore, Coventry, CV8 3LG. Contact Sally Furness on tel: (02476) 303517.

· **Keveral Farm,** St Martins-by-Looe, Cornwall, PL13 1PA. Contact Gina Cooper on tel: (01503) 250215.

· Lower Shaw Farm, Old Shaw Lane, Swindon, Wilts, SN5 9PJ. Contact Andrea Hirsch on tel: (01793) 771080.

· **Middlewood Centre of Environmental Excellence,** Roeburndale West, Wray, Lancaster, LA2 8QX. Contact Rod Everett on tel: (015242) 21880 .

· **Naturewise,** c/o Crouch Hill Recreation Centre, 83 Crouch Hill, London, N8 9EG. Contact Alpai Torgut on tel: (0171) 281 3765 .

· **New Barn Field Centre,** Bradford Peverell, Dorchester, Dorset, DT2 9SD. Contact Neville Dear on tel: (01305) 268865.

· **Pit Hill Community Environment Centre,** Holme Lane, Bradford, BD4 0QF. Contact Rhona Pringle on tel: (01274) 688766 .

· **Prickly Nut Wood,** c/o 69 Petworth Road, Haslemere, Surrey, GU27 3AX. Contact Ben Law on tel: (0966) 193154.

· **Ragman's Lane Farm,** Lower Lydbrook, Lydbrook, Glos, GL17 9PA. Contact Matt Dunwell on tel: (01594) 860123.

· **Redfield Study Centre,** Redfield Community, Buckingham Road, Winslow, Bucks, MK18 3LZ. Contact Simon Pratt on tel: (01296) 714983.

· **Springfields Community Gardens,** Sterling Crescent, Holmewood, Bradford, BD4 0DA. Contact Chris Mackenzie-Davey on tel: (01274) 688359.

· **Turners' Field,** Compton Dundon, Somerton, Somerset, TA11 6 PT. Contact Ann Morgan on tel: (01458) 442192.

Scotland

· **Earthward,** Tweed Horizons, Newtown St Boswells, Roxburghshire, TD6 0SG. Contact Dr Nancy Woodhead on tel: (01835) 822122.

· **Highland Ecology Centre,** The Old Inn, Strathconon, Muir of Ord, Ross-shire, IV6 7QQ. Contact Jim or Jo Monahan on tel: (01997) 477 260.

Wales

· **Centre for Alternative Technology.** Contact Ann MacGarry (day visits) or Joan Randle (residential) on tel: (01654) 703743.

· **Chicken Shack Housing Co-op,** Bryn Llwyn, Twyn, Wales, LL36 9NH. Contact Steve on tel: (01654) 711655.

· **Dyfed Permaculture Farm Trust,** Bach y Gwyddil, Cwmpencraig, Llandysul, Carmarthenshire, SA44 5HX. Contact Sian McNally on tel: (01559) 370438.

· **Tir Penhros Isaf,** Hermon, Llanfachreth, Dolgellau, Gwynedd, LL40 2LL. Contact Chris Dixon on tel: (01341) 440256. Email: mawddach@gn.apc.org

Ireland

· **Milleen,** Coomhola, Bantry, County Cork. Contact Mike Holden on tel: (00 353) 2766328.

· **The Ark Pc Project,** Clones, County Monaghan, Ireland. Contact Marcus McCabe & Kate Mullaney on tel: (00 353) 47 52049.

· 998 Crumlin Road, Belfast BT14 8HF. Contact Phillip Allen on tel: (02890) 716200 .

Frequently referred to texts

The following texts are essential reading for any permaculture teacher:

The Permaculture Plot, Pratt, S (Ed), Permanent Publications, 1997

Permaculture Way, Bell, G, Thorsons, 1992

The Permaculture Garden, Bell, G, Thorsons, 1994

The Basics of Permaculture Design, Mars, R, Candlelight Trust, 1996

Earth User's Guide to Permaculture, Morrow, R, Kangaroo Press, 1993

Urban Permaculture, Watkins, D, Permanent Publications, 1993

Hepburn Permaculture Gardens, 10 years of sustainable living, Holmgren, D, Holmgren Design Services, 1996

Permaculture One, Mollison, B and Holmgren, D, Tagari, 1978

Introduction to Permaculture, Mollison, B, & Slay, R M, Tagari, 1991

Permaculture: A Designers' Manual, Mollison, B, Tagari, 1988

Permaculture in a Nutshell, Whitefield, P, Permanent Publications, 1992

How to Make a Forest Garden, Whitefield, P, Permanent Publications, 1996

The History of the Countryside, Rackham, O, Dent, 1986

The Ecology of Urban Habitats, Gilbert, O, Chapman & Hall, 1991

The Wild Flowers of Britain and Northern Europe, Fitter, R , Fitter, A & Blamey, M, Collins, 1975

Serious Creativity, de Bono, E, Harper Collins, 1992

The Mind Map Book, Buzan, T & Buzan, B, BBC Books, 1995

The Web of Life, Capra, F, Anchor Books, Doubleday, 1996

Local Harvest, delicious ways to save the planet, de Selincourt, K, Lawrence & Wishart, 1997

Short Circuit, strengthening local economies for security in an unstable world, Douthwaite, R, Lilliput, 1996

Food for Free, Mabey, R, Collins, 1972

The One-Straw Revolution, Fukuoka, M, Rodale, 1978

Living in the Environment, Miller, G T, Wadsworth, 1996

Water for every farm, Yeomans, K, Keyline Designs, 1993

Our Ecological Footprint, Reducing Human Impact On The Earth, Wackernagel, M & Rees, W, New Society Publishers, 1996

A Pattern Language, Alexander, C, et al, Oxford University Press, 1977

Designing with Nature, McHarg, I, Doubleday, 1991

Designing and Maintaining Your Edible Landscape Naturally, Kourik, R, Metamorphic, 1986

The Ecology of Commerce, Hawken, P, Harper Business, 1993

Starting with a Seed... a Guide to Community Projects in the Environment, Warburton, D, WWF, 1995

Local Agenda 21 Case Studies, **Local Government Management Board (now IDA)**, 1997 onwards

Local Agenda 21 Cookbook, **Local Government Management Board (now IDA),** 1997. Both available from IDA, Layden House, 76-86 Turnmill St, London EC1M 5QU. Tel: (020) 7296 6600

Other permaculture teaching books

The Manual for Teaching Permaculture Creatively, Skye & Clayfield, R, Earthcare Education, 1991

Earth User's Guide to Permaculture – Teachers Notes, Morrow, R, Kangaroo Press, 1998

Permaculture Design Course Handbook, Sobol, G, Self-published, 1985. To obtain a copy write to: George Sobol, Jolly Lane Cottage, Hexworthy, Yelverton, Devon, PL20 6DS. Tel: (01364) 631 333

Books for trainers and educators

The Big Book of Team Building Games: Trust Building Activities, Team Spirit Exercises and Other Fun Things to do, Newstrom, J, Scannel, E, McGraw Hill, 1998

100 Training Games, Kroenhart, G, McGraw Hill, 1992

Games Trainers Play, Scannel, E, Newstrom, WJ, McGraw Hill, 1989

50 One Minute Tips for Trainers: A Quick and Easy Guide, Daele, AV, Paris, J, Crisp Publications, 1996

Basic Training for Trainers – A handbook for new trainers, Kroenhart, G, McGraw Hill, 1995

The Learning Revolution, Dryden, G & Vos, J, Accelerated Learning, 1996. Best practice from around the world

Participatory Learning and Action. A Trainers Guide, Pretty, J et al, IIED, 1995

Outdoors with Young People: a leaders guide to outdoor activities, the environment and sustainability, Cooper, G, Russel House Publishing, 1998

The New Youth Games Book, Dearling, A & Armstrong, H, Russel House Publishing, 1998

Videos

Ancient Futures, Page, J, International Society of Ecology and Culture, 1993. The follow up *Local Futures* is also excellent and concentrates on the solutions to the problems presented in Ancient Futures

The Permaculture Concept – In Grave Danger of Falling Food, Australian Broadcasting Corporation, 1992

Global Gardener – Gardening the world back to life, with Bill Mollison, Australian Broadcasting Corporation, 1991

Forest Gardening, Iota Pictures in association with Green Earth Books, 1995

Permaculture in Practice, Iota Pictures, 1997

Spirit of the Trees, available from Beckmann Home Video, PO Box 44, Leatherhead, Surrey, KT22 7AE. Tel: (01372) 805000

Sources of case study material

Local Agenda 21 Case Studies, **Local Government Management Board (now IDA),** 1997 onwards

The Permaculture Project Network has a directory of active UK projects. Contact the Association for details

See also the periodicals listed below

Periodicals

Permaculture Works. Newsletter sent free to members of the Permaculture Association (Britain)

Permaculture Magazine. High-quality, informative magazine

Permaculture International Journal (Australia), *The Permaculture Activist* (USA). Magazines with much of interest to British readers. Both available from Permanent Publications

Book & video suppliers

WWF-UK, Education Distribution, PO Box 963, Slough, SL2 3RS. Tel: (01753) 643104. Fax: (01753) 646553. Ask for the Resources and Services for Teachers Education Catalogue and the Resources & Services for Community & Business Catalogue.

Permanent Publications, Hyden House Ltd., The Sustainability Centre, East Meon, Hants, GU32 1HR. Tel: (01730) 823311. Fax: (01730) 823322. Email: hello@permaculture.co.uk Magazine, book publishing, Earth Repair Catalogue mail order service, and the Permaculture Magazine Information Service website.

éco-logic books, 19 Maple Grove, Bath, BA2 3AF. Tel: (01225) 484472. Fax: (0117) 942 0164. Mail order catalogue.

Worldly Goods, 10-12 Picton Street, Bristol BS6 5QA. Tel: (0117) 942 0165. Fax: (0117) 942 0164. Wholesale distribution to shops, environment centres, permaculture groups and teachers.

Centre for Alternative Technology, Machynlleth, Powys, SY20 9AZ. Tel: (01654) 703409.

Intermediate Technology Publications, 103-105 Southampton Row, London WC1B 4HH.

Henry Doubleday Research Association, Ryton-on-Dunsmore, Coventry, CV8 3LG. Tel: (01203) 303517.

British Trust for Conservation Volunteers, 36 St Marys Street, Wallingford, Oxon, OX10 0EU.

Schumacher Book Service, Ford House, Hartland, Bideford, Devon, EX39 6EE. Tel: (01237) 441 621. Fax: (01237) 441203.

Wholistic Research Company, Bright Haven, Robin's Lane, Lolworth, Cambridge, CB3 8HH. Tel: (01954) 781074.

Green Books, Foxhole, Dartington, Totnes, Devon, TQ9 6EB. Tel: (01803) 863260.

Beckmann Home Video, PO Box 44, Leatherhead, Surrey, KT22 7AE. Tel: (01372) 805000.

Reading International Solidarity Centre (RISC) have a wide range of Development Education materials, including teachers' packs and videos. All available through their mail order catalogue. Contact RISC at 35-39 London Street, Reading, Berks, RG1 4PS. Tel: (0118) 958 6692.

The Internet

If you are new to the net, or just want to hone your skills, the following titles may be of use:

Search Engines for the World Wide Web: Visual Quickstart Series, Glosebrenner, A, Peachpit Press, 1997

Internet Research Companion, McKin, GW, Que Education & Training, 1997

The On-Line Research Handbook: Concise Guides Series, Mead, H & Clark, A, Berkeley Publishing Group, 1997

Web Search Strategies, Pfaffenberger, B, IDG Books Worldwide, 1996

Useful starting points

The Permaculture Association is building an extensive range of links relating to the different aspects of permaculture. Visit the website for current links.

➪ *Contributors*

Graham Bell is the author of The Permaculture Way and The Permaculture Garden, and has been teaching permaculture nationally and internationally for 10 years.

Phil Corbett has worked with organic horticulture and ecology for many years. He started the Own Root Fruit Tree project and is the founder of the Cool Temperate Nursery.

Chris Dixon is a published author, wilderness regenerator, holder of the Diploma in Permaculture Design, and has served three years on the Council of the Permaculture Association (Britain), one year as Chairman. He teaches performing arts, is developing a PC holding in the Snowdonia National Park, speaks Cymraeg (rhugl) and has been known to play the accordion but prefers an analogue Korg.

Lyn Dixon has over 30 years experience of working with horses, and holds the Monty Roberts Preliminary Certificate in Horsemanship. Lyn has also worked on farms with cattle and sheep for over 20 years, at home has managed goats, iron age pigs, sheep, ducks, geese, hens and a cow. She takes horses for join-up, starting and remedial work. Interests and skills include animal psychology, complimentary therapies, growing organic veg for veggie box scheme and pursuing her training with Monty Roberts.

Micah Duckworth. Inputs and influences: Australian forests, Springfield Community Gardens, nature, friends and family, Schauberger, Tesla, Stanshal, Taylors Landlord. Outputs and outpourings: teaching permaculture, water and appropriate technology research, IT , kites, gizmos and music.

Matt Dunwell has been developing Ragman's Lane Farm for the last seven years. The farm produces beef, lamb, pork, chicken, eggs, honey, charcoal, cider, shii-take mushrooms, and vegetables through a veg box scheme.

Rod Everett set up the Middlewood Charitable Trust, a 200-acre plot with demonstration gardens, low energy buildings and native woodland. He is currently developing an organic advisory service.

Mike Feingold. Born into a farming family in East Africa, Mike is experienced in extablishing self-help community organisations and community action. An amateur with a passion for the regeneration of degraded systems mainly in tropical regions, as well as work in Britain, he has been active in teaching and promoting sustainable living for the last 25 years.

Mark Fisher is an organic gardener who grew into permaculture.

Patsy Garrard has a background in housing, social policy research, tree work and organic horticulture. She launched into permaculture after hearing Bill Mollison speak in 1989. In the early 1990s Patsy helped run the Association's national office with George Sobol and they co-produced the bulletin Is Anybody Out There? As well as teaching permaculture in Britain and Central/Eastern Europe, Patsy also trains new teachers so that she can spend more time on land-based and publishing projects.

Andrew Goldring is a reformed art student who now spends his time teaching and designing.

Judith Hanna is policy co-ordinator at the National Centre for Volunteering. Her experience of environmental and community development work includes serving as Chair and Secretary of the Permaculture Association (Britain), Chair and Trustee of the New Economics Foundation, on the founding Board of LETSlink London, Assistant Director of Transport 2000, editing Local Transport Today, and parliamentary worker with the Campaign for Nuclear Disarmament. Raised on an Australian farm, she took her Design Course with Naturewise in London in 1994, and is involved in a range of local activities including LA21.

Maddy Harland is the editor of Permaculture Magazine, Solutions for Sustainable Living and a director of Permanent Publications, an independent publishing company specialising in publishing and distributing information about permaculture design and other approaches to sustainable living. Maddy is also part of a team working to turn a redundant military base into a living and working Sustainability Centre in Hampshire, England.

Peter Harper is a biologist, gardener and landscape designer who works at the Centre for Alternative Technology. He writes widely on environmental themes and is the author of Radical Technology, The Natural Garden Book, and Fertile Waste. He also runs 'The Lifestyle Lab', an experimental urban dwelling where he lives with his long-suffering family. He is mildly obsessed by compost.

[Jane Hera?]

Andy Langford is currently Director of the Permaculture Academy, and has been closely involved in the development of the Association since its beginning. Recent projects include design input to the Springfields Community Garden project in Bradford, design of the Blackdown Hills Sustainable Development Initiative for Devon County Council and the design of an innovative low-cost, distributed learning network for permaculture designers. Andy now works for a wide range of clients across the UK including local government, and is currently developing strategies to aid community participation in Local Agenda 21.

Ben Law is a permaculture teacher and designer, specialising in woodland systems. He lives and works in Prickly Nut Wood.

Chris Mackenzie-Davey, has been a permaculture designer since 1995. Previously trained in horticulture and landscape design, Chris has been responsible for establishing the Springfield Community Garden in Bradford.

Anne-Marie Mayer has a background in human nutrition and health. She is currently investigating the connections between the nutritional quality of food and agriculture.

Dave Melling lives and works in the District of Bradford. He has been involved in a wide range of community based activities for the last 20 years. Dave works as a Local Agenda 21 Officer and has been closely connected to Springfield Community Garden since its inception.

Oak has been involved in woodland and conservation work for the last 15 years. A member of the Keveral Farm community, Oak helps to convene courses and is active in the ongoing implementation of the design for the site.

Simon Pratt has been teaching permaculture since October 1989, mainly from his home at the Redfield Community in Buckinghamshire. As well as being an active member of the Permaculture Association, he has produced three editions of The Permaculture Plot. His main current interest is in ecological building and he has designed and built three straw bale buildings at Redfield.

Maryjane Preece enjoys singing, spreading hope and sowing seeds.

Thomas Remiarz. Home: my body, then Europe with all the bits, then Cornerstone Housing Co-op, Leeds. Likes: good food, good friends, good music, coming home and going away. Dislikes: capitalism, patriachy and all other forms of oppression. Does: permaculture design (requested or not), education (ditto), home improvements and various campaigns.

Jamie Saunders is a husband, father, 'whiteboard fiend' and LA21 consultant for Bradford Council.

Skye is currently living and working in Mexico. He co-wrote the book The Manual for Teaching Permaculture Creatively and was a member of the Crystal Waters village in Australia.

George Sobol had a previous life as a furniture designer and maker. He convened and attended his first Design Course taught by Andy Langford in 1989/90. He was the national co-ordinator for the Permaculture Association for four years, and spent another four years co-producing the bulletin Is Anybody Out There? He is currently working on the merging of permaculture with his mother-tongue (Russian) and on making permaculture more visible.

Angus Soutar, world citizen with bioregional agendas, making connections, self-employed (no job too large, no job too small).

Krysia Soutar is helping to create a peaceful world by working with Nature.

Bryn Thomas is active in education, community projects, LETS, garden/farm design, agroforestry, woodland & wildlife. A former RSPB reserve warden, Bryn holds the diploma in Permaculture Design and a HND in Conservation Management.

Joanne Tippett set up the Holocene Design Company, which offers design and education services in strategic planning for sustainability (The Natural Step), permaculture, ecological architecture and site planning.

Mark Warner is an environmental generalist based in Bristol where he is currently developing a bioregional project for the River Avon catchment area. Mark has been involved with permaculture for 10 years and launched the Ecovillage Network, UK in 1996 following up from an ecovillage design course with Max Lindegger.

Cathy Whitefield, Cert Couns, Lic Ac, has wide experience as a group facilitator and teacher. She has a background in counselling, complementary medicine, yoga and spiritual practice. She believes that healing ourselves is an intrinsic part of healing the Earth.

Patrick Whitefield is a permaculture teacher, writer and consultant. He originally qualified in agriculture, and his experience includes farming, gardening, nature conservation, crafts and green politics. He is the author of Permaculture in a Nutshell, and How to Make a Forest Garden.

Nancy Woodhead. When asked what tree she would be, if re-incarnated with the same qualities she has now, Nancy chose Elder, as it is 'very useful and likely to pop up anywhere'; she adds 'teaching and demonstrating permaculture practice is a bit like falling off a log, landing on your feet gets easier after the first few times.'

Contact details of contributors can be obtained from the Permaculture Association.

➡ *Acknowledgements*

Managing Editor

Andrew Goldring

Associate Editor

Patrick Whitefield

Further editing support was provided by Martine Drake, Judith Hanna, Angus Soutar, Maddy Harland, Mike Feingold and George Sobol.

List of contributors

Graham Bell, Phil Corbett, Chris Dixon, Lyn Dixon, Micah Duckworth, Matt Dunwell, Rod Everett, Mike Feingold, Mark Fisher, Patsy Garrard, Andrew Goldring, Judith Hanna, Maddy Harland, Peter Harper, Jane Hera, Andy Langford, Ben Law, Chris Mackenzie-Davey, Anne-Marie Mayer, Dave Melling, Oak, Simon Pratt, Maryjane Preece, Thomas Remiarz, Jamie Saunders, Skye, George Sobol, Angus Soutar, Krysia Soutar, Bryn Thomas, Joanne Tippett, Mark Warner, Cathy Whitefield, Patrick Whitefield, Nancy Woodhead. Short biographies of each contributor can be found on pages 373-375.

With help from

Peter Andrews, Sue Cameron, Liam Egerton, Tim Harland, Philip Hinton, Marcus McCabe, Kate Mullaney, Julie, Florence and Gabriel Nutchey, Lin Simonon, Rob Squires and Wolf White.

The Permaculture Association would like to thank Ken Webster, Cherry Duggan, Christine Stone and the design team at WWF-UK, for all their work in nurturing this project from start to finish.

Photo credits and captions

p16 top - Coffee breaks give students a chance to get to know each other; bottom - The world is our classroom! © Graham Bell

P17 Briefing students prior to a tour of the gardens at Tweed Horizons. © Graham Bell

p18 Dancing and other celebrations help to make the course special. © Graham Bell

p22 Terminally boring teaching material! © Patrick Whitefield

P23 top - Venues and visits should be accessible to all; bottom - Design exercises give teachers the opportunity to gauge students' understanding. © Patrick Whitefield

P24 Practical group task - making an island out of car tyres. © Patrick Whitefield

P29 Mind maps mimic the way the brain stores and integrates knowledge. © Joanne Tippett

P30 Uncovering people's knowledge using visual techniques. © Joanne Tippett

p80 Multi-functional elements. © Joanne Tippett

p8l Using three dimensional space. © Joanne Tippett

p85 Permaculture principles. © Joanne Tippett

p87 Making sectors outside. © Joanne Tippett

P92 The importance of 'cycling'. © Joanne Tippett

Plo3 Presenting a design as a clay model. © Tim Bush

P107 le{t- Making an A-frame. © Phillip Hinton; top right- Using an A-frame. © Anne-Marie Mayer; bottom right - Base map exercise: surveying the site. © Patrick Whitefield

plo8 Students working on the Input/Output exercise. © Andrew Goldring

P134 Base map exercise: surveying the site. © Patrick Whitefield

P135 Design questionnaire: asking questions on the site. © Patrick Whitefield

p136 top - Base map exercise: drawing up the map. © Patrick Whitefield; middle - Students
    discussing design strategies; and bottom - Students discussing their designs.
    © Andrew Goldring

P144 Observation is a critical skill and many opportunities should be created to develop students'
    abilities. © Andrew Goldring

P148 Listening to the landscape. © Patrick Whitefield

P1S4 Unused assets. © Graham Bell

P202 Ken Fern from 'Plants for a Future' showing students around their huge collection of edible
    perennials. © Anne-Marie Mayer

p210 Soil practicals -left- Digging an inspection pit; top right- Examining a sample on the desktop;
    bottom right- Examining a sample in the field. © Patrick Whitefield

p212 Students participate in the 'Web of Life'. © Anne-Marie Mayer

p217 Using a tray of sand to show contours and water strategies. © Patrick Whitefield

p224 Herb spirals can be made easily and quickly. © Andrew Goldring

P231 Ragmans Lane Farm - top - Preparing to make charcoal; middle - No smoke without fire?;
    bottom - Green woodworking. © Patrick Whitefield

P232 Chairs made by students, Ragmans Lane Farm. © Patrick Whitefield

P243 Site visits reinforce learning. An organic greenhouse business in Iceland. © Graham Bell

P2So top and bottom - Robert Hart and his forest garden. © Patrick Whitefield

P2S6 A visit to the Agro Forestry Research Trust. Martin Crawford explains the current experiments.
© Patrick Whitefield

P2S8 Matt Dunwell showing students the pond at Ragmans Lane Farm.

p266 Tour of the vegetable box scheme at Ragmans Lane Farm. © Phillip Hinton

p282 Straw bale building on a course at the Redfield Community. Activities such as this give students
    a great feeling of achievement. © Simon Pratt

P307 Feeling earth energies. A guided visualisation. © Graham Bell

⇨ **About the partners**

About the Permaculture Association

The Permaculture Association (Britain) is an education and research charity that supports individuals, groups and projects to design and create permaculture systems across Britain. We are part of an international network that works in over 100 countries and share the belief that the health and wellbeing of the earth and its inhabitants depends on liberating the skills, creativity and wisdom of the Earth's human population.

For more detailed information about the Association contact:
BCM Permaculture Association
London WC1N 3XX.
Tel/Fax: 0845 4581805
Email: office@permaculture.org.uk
Web site address: www.permaculture.org.uk

About Permanent Publications

Permanent Publications is a small independent publishing company which started life in 1990 in the spare bedroom of Tim and Maddy Harland's cottage. Their mission was to publish practical books to help people live more healthy and self-reliant lives. Permaculture design, a synthesis of useful practices, such as low maintenance organic gardening, alternative technology, ecological house design, sustainable agriculture and community economics, quickly became the central focus of the company.

Permanent Publications is now based at the Centre for Sustainability, where a team of eight committed people publishes and distributes hundreds of books and videos world wide through the Earth Repair Catalogue on everything from mulching to mediation and solar heating to spirituality. It also publishes Permaculture Magazine - Solutions For Sustainable Living and has an information service and full catalogue on their website.

These are exciting times with sustainability becoming recognised as a key focus for the next Millennium. Whatever direction the company chooses to take in the future, the original mission remains - to encourage good health, for people and the planet, and to inspire a spirit of interdependence and self-reliance.

For further information, contact:
Permanent Publications
The Sustainability Centre, East Meon, Hampshire GU32 1HR.
Tel: (01730) 823311.
Fax: (01730) 823322.
Email: all@permaculture.co.uk
Web site address: www.permaculture.co.uk

About WWF

WWF is the world's largest global environment network, with national offices and affiliate organisations active in 96 countries.

WWF aims to conserve nature and ecological processes for the benefit of all life on Earth. By stopping, and eventually reversing the degredation of our natural environment, we strive for a future in which people and nature can live in balance.

This mission can only be achieved if people recognise and accept the need for sustainable, just and careful use of natural resources. WWF-U K believes that education has a key role to play in this process. We are therefore working with schools, colleges, further and higher education, with community groups, and with business and industry. Our comprehensive environmental education programme includes resource development, ICT projects, curriculum development, professional and vocational training, business toolkits and work with local authorities.

If you would like further details about WWF-U K's education programme, please write to:

WWF-UK
Education and Awareness
Panda House, Weyside Park, Godalming, Surrey GU7 1XR.
Tel: (01483) 426444.
Fax: (01483) 426409.
Email: wwf-uk-ed@wwfnet.org
Web site address: www.wwf-uk.org